Ali Mohammed Mooshly Al-Bahadly

Proteins

Ali Mohammed Mooshly Al-Bahadly

Proteins

Noor Publishing

Imprint
Any brand names and product names mentioned in this book are subject to trademark, brand or patent protection and are trademarks or registered trademarks of their respective holders. The use of brand names, product names, common names, trade names, product descriptions etc. even without a particular marking in this work is in no way to be construed to mean that such names may be regarded as unrestricted in respect of trademark and brand protection legislation and could thus be used by anyone.

Cover image: www.ingimage.com

Publisher:
Noor Publishing
is a trademark of
Dodo Books Indian Ocean Ltd. and OmniScriptum S.R.L publishing group

120 High Road, East Finchley, London, N2 9ED, United Kingdom
Str. Armeneasca 28/1, office 1, Chisinau MD-2012, Republic of Moldova, Europe
Managing Directors: Ieva Konstantinova, Victoria Ursu
info@omniscriptum.com

Printed at: see last page
ISBN: 978-620-9-34125-0

Proteins

Ali Mohammed Mooshly Al-Bahadly

PhD BIOCHEMISTRY

Ja'afar Al-Sadiq University

Dedication

To those unmatched by anyone in the universe, to whom God has commanded us to be dutiful, and to those who gave endlessly and offered what can never be repaid—to you, my beloved mother and father, I dedicate this book.

Your love, sacrifices, and unwavering support have guided me through every step of my academic journey.

You have been my greatest inspiration, and I offer this work to you with heartfelt gratitude for all you have given me throughout my life.

Introduction

Science is a nation's instrument of advancement, and humanity cannot progress without it. Therefore, I have dedicated my efforts to writing and completing this book in a manner befitting its importance to all people.

The aim of this book is to provide an authoritative, highly focused, science-based, and time-efficient resource.

This book is divided into five chapters, each addressing a specific topic related to proteins.

I hope that I have succeeded in preparing this book and that it will be well received by readers.

Ali Mohammed Mooshly Al-Bahadly

PhD BIOCHEMISTRY

Ja'afar Al-Sadiq University

Contents

Chapter One

Introduction to Proteins

Proteins

Proteins are polymers of amino acids linked by alpha peptide bonds, and they can be represented by primary, secondary, tertiary, and even quaternary structures. There are 20 common amino acids in biochemistry, and proteins function as structural support, enzymes, hormones, biochemical catalysts, building blocks, as well as catalysts for cell death .

Proteins make up more than half of the solid matter in the body, and they are responsible for almost all the reactions that occur within the tissues, in addition to their structural importance .

Proteins are large molecules with colloidal properties , However, their molecular weights are not necessarily large, and their molecular weights range from a random minimum of 10,000 to several million , based on their nucleic acid content, the human body contains approximately 100,000 different types of proteins.

Proteins are usually unique to the species from which they are derived, as there is a difference between human serum albumin and serum albumin of duck, dog, or chicken. Although they are similar in function, they differ slightly in composition and are immunologically distinct.
Humans consume proteins, which are digested in the digestive system and absorbed in the form of small peptides (dipeptides and tripeptides) and free amino acids, then used to resynthesize proteins in the cells of the human body.
Also, some amino acids are used in the synthesis of specific (non-protein) products such as nucleotides, hormones, neurotransmitters, nitric oxide, creatine, glutathione, glucosamine, and other factors.
More clearly, proteins can be defined by their four structural levels: primary, secondary, tertiary, and quaternary.

The primary level is the primary structure, which is the simplest protein structure, consisting of the linear arrangement of amino acid residues. All residues are connected by peptide bonds. These bonds have specific sites for alpha, beta, and gamma carbon atoms that correspond to specific positions relative to the peptide bond. This structure is also called the protein backbone.

The second level of protein structure is the secondary structure, which consists of various shapes formed through hydrogen bonds, including alpha helices, beta pleated sheets, and beta turns.

The third level of protein structure is the tertiary structure. This structure consists of the three-dimensional shape formed when the backbone of the polypeptide chain interacts with an aqueous environment, which begins to form immediately after the newly synthesized polypeptide chain exits the terminal end of the ribosomal subunit complex. The polypeptide chain isolates the hydrophobic residues and exposes the hydrophilic residues to achieve thermodynamic stability.

The fourth and final level of protein structure is the quaternary structure. This level occurs when complexes of polypeptide chains, called subunits, are formed. An example of this is hemoglobin, whose quaternary structure forms when chemical interactions link two alpha subunits and two beta subunits together. Therefore, it is appropriate to say that the quaternary structure is the three-dimensional arrangement of two or more polypeptides in the protein, each of which folds independently.

Proteins contain disulfide bonds, and these bonds are capable of accepting electrons from alcohol groups. This concerning issue manifests through hormonal degradation. Hormones are known for being proteins rich in disulfide bonds, so when hormones are placed in an aqueous environment rich in alcohol groups, they degrade and lose their biological functions. It appears that this condition is the reason for the halted growth of adolescents who excessively consume alcohol (due to growth hormone degradation), as well as the reason why breastfeeding mothers with alcohol use disorders stop breastfeeding (due to degraded prolactin), and also the reason why alcohol addicts develop diabetes over time (due to insulin degradation).

Proteins were first described by the Dutch chemist Gerardus Johannes Mulder and the Swedish chemist Jöns Jacob Berzelius in 1838 proposed the term " protein "to describe these molecules, the word protein is derived from the Greek word πρώτειος (prōteîos) meaning "first" and the addition (in) which appends the end of various chemical compounds, Gerardus Johannes Mulder conducted initial tests on common proteins and found that most proteins have almost the same empirical formula, leading him to the incorrect conclusion that proteins could be composed of a single type of molecule . However, in 1902, Franz Hofmeister and Hermann Emil Fischer worked on understanding that proteins are polypeptides. In 1926, James Sumner demonstrated that the enzyme urease is indeed a protein, thereby proving the fundamental role of proteins as enzymes.

Insulin is also considered to be the first protein whose sequence was determined by Frederick Sanger in 1955, where Sanger correctly identified the amino acid sequence of insulin and was awarded the Nobel Prize in Chemistry for his work on the structure of proteins in 1958 .

The physiological role of proteins and their functions in the human body

Proteins are macronutrients that support the growth and maintenance of body tissues, and amino acids are the basic building blocks of proteins, classified into essential, non-essential, and semi-essential.

Non-essential amino acids are naturally produced in the body, while essential amino acids are obtained from protein-rich foods such as meat, poultry, and legumes. Therefore, the human body should get between 10% and 25% of its daily caloric needs from proteins.

Proteins perform most of the work in cells and are essential for the structure, function, and regulation of the body's tissues and organs. Proteins are made up of hundreds or thousands of smaller units called amino acids, which are linked together in long chains. There are 20 different types of amino acids that can be combined to form a single protein. Physiologically, proteins are large molecules composed of amino acid subunits called monomers. These amino acids are covalently bonded to each other to form long chains called polypeptides, which then fold to form a specific three-dimensional shape.

Folded polypeptide chains are sometimes functional by themselves , and sometimes they may combine with additional polypeptide chains to form the final protein structure, and often non-polypeptide groups are also necessary in the final protein , for example, the blood protein hemoglobin consists of four polypeptide chains, each of which contains a heme molecule, which is A ring structure with an iron atom in the middle .

And some of the functions of proteins in the body are:

- Proteins are considered important sources of energy.
- Proteins work on the growth and building of body tissues.
- Regenerating and replacing damaged tissues.

- Proteins are involved in the composition of enzymes, hormones, and antibodies.
- Proteins maintain water balance in the body.
- Proteins help maintain the acid-base balance in the body.

The chemical composition of proteins

We previously mentioned that there are 20 types of amino acids in proteins, each with different chemical properties. A protein molecule consists of a long chain of these amino acids, each linked to its neighbor through a covalent peptide bond, as shown in the following figure.

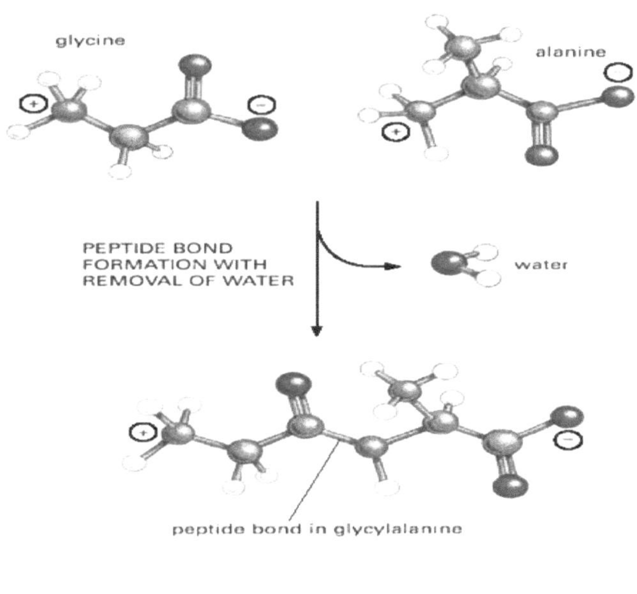

Peptide bond formation

The repeated sequence of atoms along the nucleus of a polypeptide chain is called the (polypeptide skeleton).

This repetitive sequence is associated with parts of amino acids that do not participate in the formation of the peptide bond, giving each amino acid its unique properties, as shown in the figure.

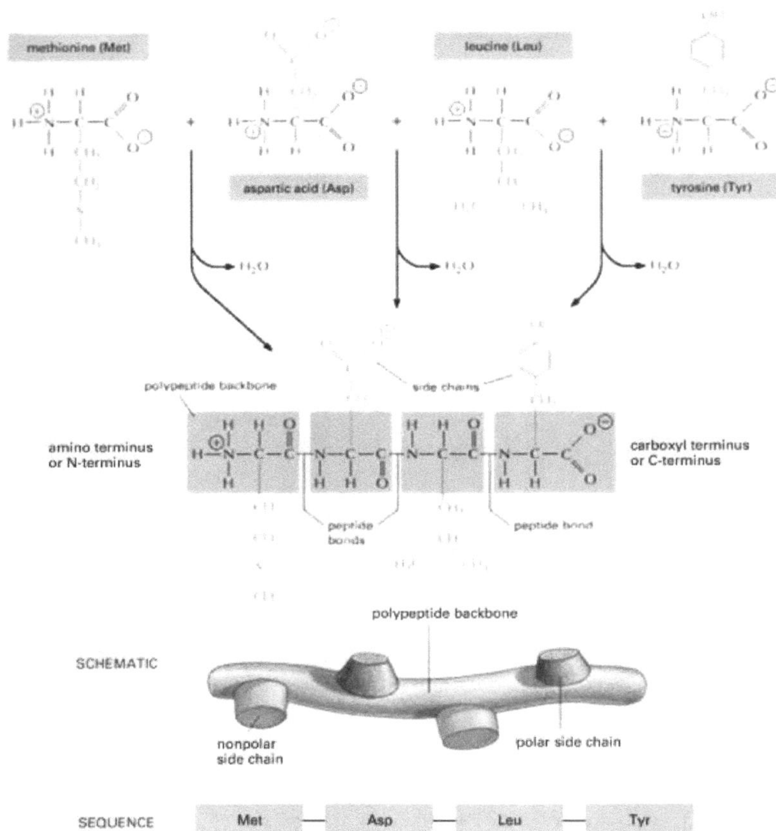

Side chains in the structure of a polypeptide chain

Some of these side chains are nonpolar and have hydrophobic properties, while others carry negative or positive charges, and there are also reactive chains, and so on.

Protein consists of a polypeptide structure with attached side chains, each type of protein differs in its sequence and number of amino acids. Therefore, the sequence of chemically different side chains is what distinguishes each protein, the two ends of the polypeptide chain differ chemically: the end carrying the free amino group (NH_3^+, also written as NH_2) is the amino terminus, and the end carrying the free carboxyl group (COO^-, also written as COOH) is the carboxyl terminus, the amino acid sequence of a protein is always presented in the N to C direction, from left to right.

Amino acids

Amino acids are the basic units that make up proteins, each amino acid has the same basic structure, which consists of a central carbon atom or alpha carbon (α), bonded to an amino group (NH2), a carboxyl group (COOH), and a hydrogen atom. In addition, each amino acid contains another atom or group of atoms attached to the central atom known as the (R) group, as shown in the following figure.

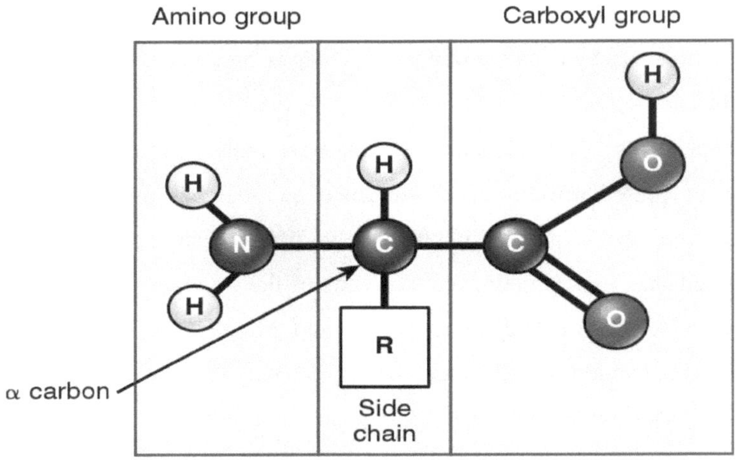

Amino acid structure

We previously mentioned that there are 20 common amino acids found in proteins. Amino acids are divided into three groups: essential, non-essential, and semi-essential. Semi-essential (conditional) amino acids are produced in the body under normal conditions, but they are classified as essential (their production becomes insufficient) during stress, injury, or illness.

Essential amino acids

Essential amino acids include nine types: leucine, isoleucine, valine, histidine, methionine, lysine, phenylalanine, tryptophan, and threonine , due to the human cells or other mammalian cells cannot synthesize them, so these amino acids must be obtained from external sources .

-Leucine (Leu), isoleucine (Ile), valine (Val): they are branched chain amino acids, have a molecular structure with a branch, these branched chain amino acids are found in abundance in muscle proteins, they stimulate muscle growth, repair damaged tissues in the body and provide energy during exercise.

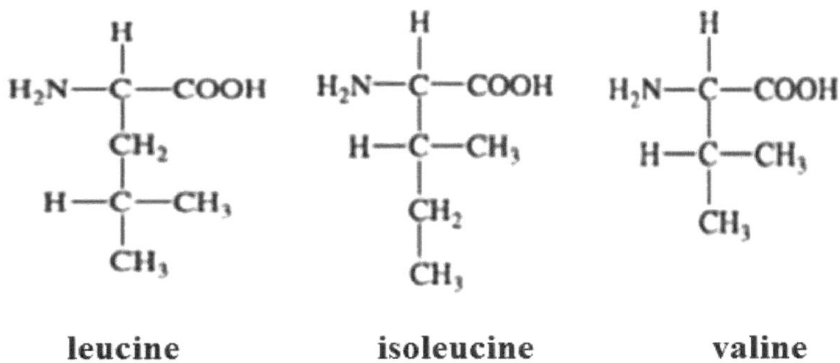

Structures of leucine , isoleucine , valine

-Lysine (Lys): it is one of the most common essential amino acids where, in addition to alarginine and histidine, it belongs to the group of basic amino acids (it has a base fraction), and the percentage of lysine is low in foods such as bread and rice, Lysine deficiency can lead to serious problems such as stunted growth and severe diseases.

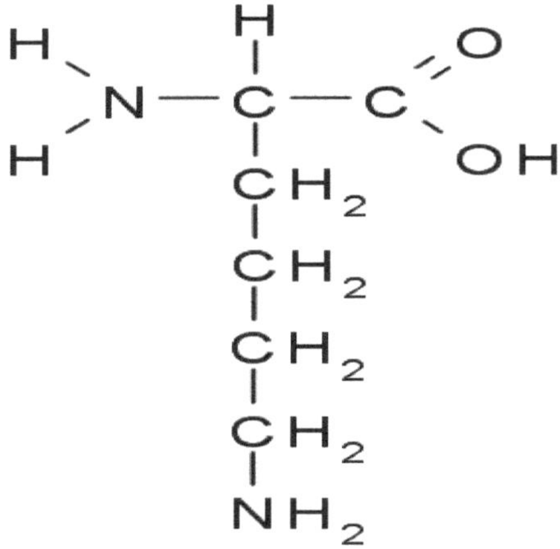

Structures of Lysine

-Threonine (Thr): it is an essential amino acid that the human body cannot synthesize, but it must be present in proteins in the diet, an adult needs about 20 milligrams of threonine per kilogram of body weight per day, as for plants and microorganisms, threonine is made from aspartic acid, and threonine is used to make the active site for enzymes.

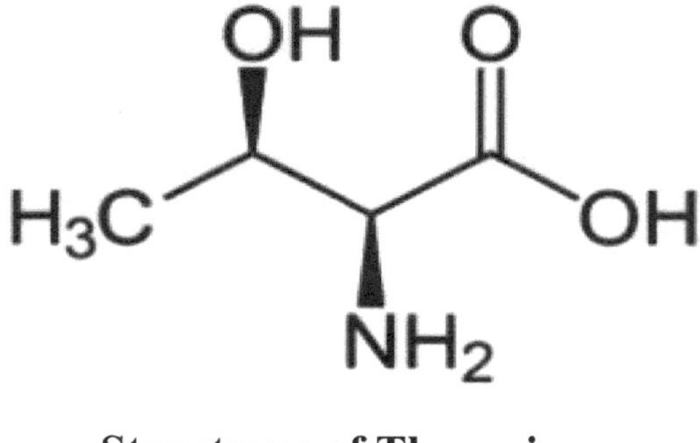

Structures of Threonine

-Phenylalanine (Phe): It is an essential, non-polar amino acid necessary for children's growth and protein metabolism in children and adults. It is found in meat, bird meat, fish, eggs, milk, and leg umes. Phenylalanine is used to make many types of beneficial amines.

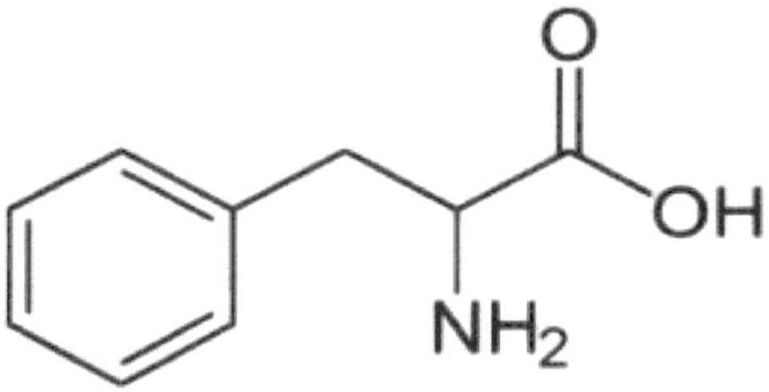

Structures of Phenylalanine

-Methionine (Met): It is an essential amino acid used to make many different substances that the body needs. Methionine plays a crucial role in the metabolism process , methionine also participates in the formation of blood vessels and various other processes related to DNA transcription, gene regulation, and gene expression. In 1921, John Howard Mueller isolated methionine for the first time.

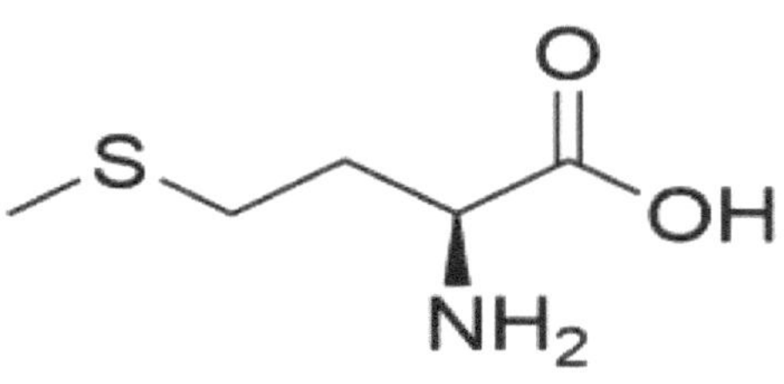

Structures of Methionine

-Histidine (His): An essential amino acid used to make histamine. It is of the alpha type where the side chain is in the form of a five-membered ring in which the nitrogen atoms form two corners of this ring. This side chain carries a positive charge on one of the nitrogen atoms in the ring, making it a positively charged amino acid and despite containing a positive charge. However , it is a weak base at neutral pH.

Hemoglobin contains a high percentage of histidine, which helps counteract the effects of excess carbon dioxide. The recommended amount of histidine for adults ranges between 0.56 and 0.84 grams per day.

Histidine was first isolated in 1896 by the scientists Albrecht Kossel and Sven Hedin.

Histidine is derived from the Greek word "histion," which means tissue, due to its first isolation from salmon fish tissues.

Histidine is considered a part of the structure and effectiveness of many hormones and receptors that affect kidney functions, neurotransmission, stomach secretion, and the immune system.

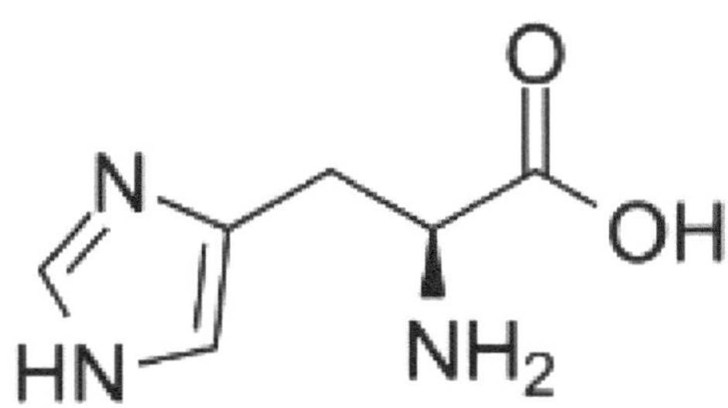

Structures of Histidine

-Tryptophan (Trp): It is an essential amino acid that the human body cannot synthesize, so it must be obtained from food. To maintain nitrogen balance in the body, an average daily dose of 3.5 mg of tryptophan per kilogram of body weight should be taken from dietary sources such as chocolate, fish, eggs, and dairy products.

A large portion of tryptophan remains in the intestines, while the rest is transported to the brain, skeletal muscles, and heart. In addition, only free plasma tryptophan can cross the blood-brain barrier via large neutral amino acid transporters, where it is then metabolized into neuroactive substances.

The level of free tryptophan in plasma depends on its binding rate to albumin, this binding is essential for the transport of tryptophan throughout the body and can be affected by albumin levels and the availability of binding sites. Free fatty acids can bind to albumin, which in turn reduces the binding rate of tryptophan and raises the levels of free tryptophan in plasma.

There are two competitive pathways for tryptophan:

1-The serotonin pathway, which produces monoamine neurotransmitters such as serotonin and melatonin, and plays a crucial role in regulating human behavior, sleepiness, and appetite.

2-The kynurenine pathway, which produces both neuroprotective metabolites (such as kynurenine and kynurenic acid) and neurotoxic metabolites (such as kynulenic acid and 3-hydroxykynurenine) at the central level.

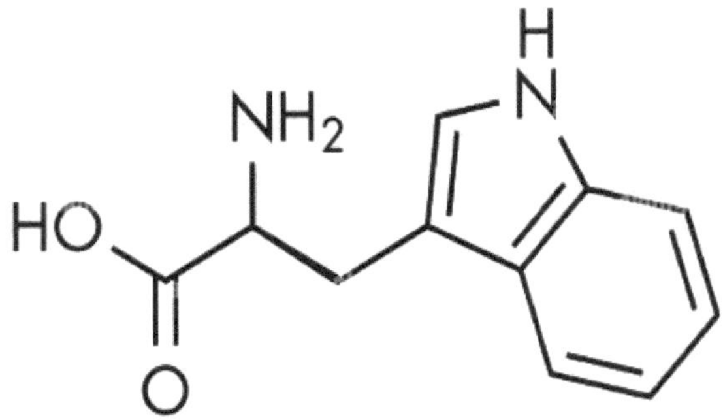

Structures of Tryptophan

Non-essential amino acids

The body naturally produces non-essential amino acids without the need to obtain them from food. These acids are essential for promoting growth, playing an important role in children during growth stages, as well as for pregnant and breastfeeding women.

Non-essential amino acids include: alanine, asparagine, aspartic acid, glutamic acid, glutamine, arginine, proline, serine, glycine, cysteine, and tyrosine.

Some non-essential amino acids are known to be conditional and are called semi-essential amino acids, meaning that under certain conditions such as illness or stress, their production becomes insufficient, so the body needs to obtain them from food. Examples of these amino acids are: glutamine, arginine, proline, serine, glycine, cysteine, and tyrosine.

-Alanine (Ala): is an alpha aliphatic non-polar amino acid due to the presence of an amino group and a carboxylic acid group, both attached to the central carbon atom which also carries a methyl side chain. It exists in two different optical isomers: (L-Alanine) and (D-Alanine). The L-isomer is incorporated into proteins, while the D-isomer is found in peptides in some bacterial cell walls, in the tissues of many crustaceans, and in some peptide antibiotics. Both forms are characterized by a sweet delicious taste when tasted in solution.

In 1850, the German scientist Adolf Strecker synthesized alanine for the first time by combining acetaldehyde and ammonia with hydrogen cyanide.

Both alanine and glutamine are major pillars in the process of glucose formation in the liver and urea formation, by converting alanine reaching the liver into pyruvate, which is used in glucose formation.

Alanine is produced in peripheral tissues through transamination processes with glutamate, branched-chain amino acids, and other amino acids, and after its release into the systemic circulation, alanine is primarily absorbed by the liver and, to a lesser extent, by the kidneys.

Alanine supports liver function, is used to break down tryptophan and vitamin B-6, and is considered an energy source for the central nervous system and muscles. It also strengthens the immune system and helps prevent hypoglycemia (low blood sugar) by producing the glucose needed by the body, additionally, alanine improves alcohol metabolism .

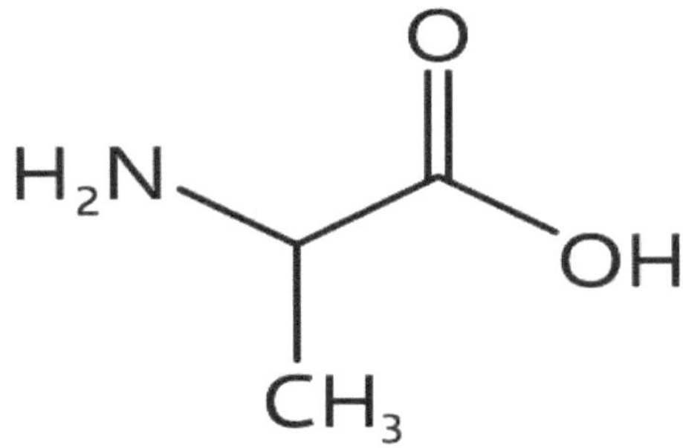

Structures of Alanine

-Asparagine (Asn): It is an alpha amino acid that contains an alpha amino group, an alpha carboxyl acid group, and a carboxamide side chain. Thus, it is a polar aliphatic amino acid. It is a non-essential amino acid, meaning the human body produces it.

There are also two sources for this acid: an animal source (poultry meat, beef, eggs, seafood, fish, dairy products) and a plant source (potatoes, nuts, legumes, soybeans, and whole grains).

Asparagine was the first amino acid to be isolated, where the French chemists Louis Nicolas Vauquelin and Pierre Jean Robiquet first isolated asparagine in crystalline form in 1806.

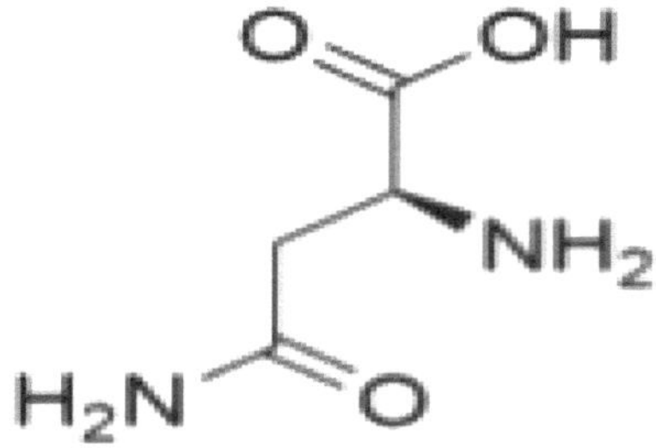

Structures of Asparagine

-Aspartic acid (Asp): is an amino acid that carries a negatively charged carboxyl group on the side chain, which gives aspartic acid an overall negative charge at physiological hydrogen ion concentrations (pH approximately 7.3).

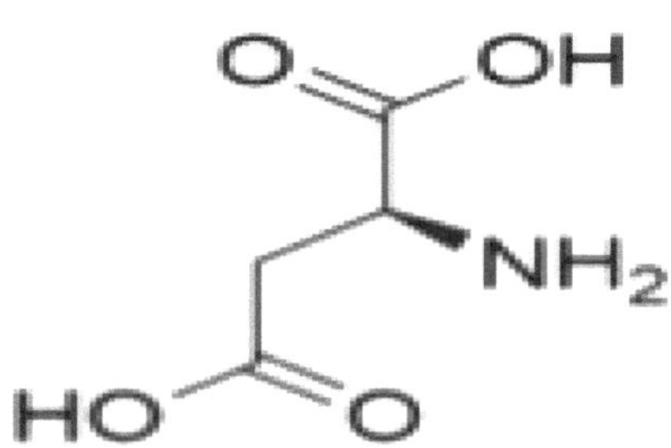

Structures of Aspartic acid

Although aspartic acid is considered a non-essential amino acid, it plays a fundamental role in metabolism by building other amino acids and biochemical compounds in the citric acid cycle. Among the biochemical compounds synthesized from aspartic acid are arginine, methionine, asparagine, threonine, lysine, isoleucine, and many nucleotides.

Aspartic acid was first discovered in 1827 by the Frenchmen Auguste-Arthur Plisson and Étienne Ossian Henry through the hydrolysis of asparagine.

Aspartic acid is considered an excitatory neurotransmitter in the brain, it is found in different brain regions. It also participates in transmitting signals between neurons, especially in the spinal cord. Additionally, aspartic acid is involved in the treatment of chronic fatigue. Aspartic acid stimulates the transfer of nicotinamide adenine dinucleotide (NADH) molecules from the main cell body to the mitochondria, where they are then used to produce adenosine triphosphate (ATP), which serves as the fuel for all cellular activities. Aspartic acid also functions to:

1-Assisting in the transport of essential minerals for the health of ribonucleic acid (RNA) and deoxyribonucleic acid (DNA) to the cells .

2-It helps maintain high concentrations of nicotinamide adenine dinucleotide (NADH) in brain cells, enhancing mental sharpness which leads to increased production of neurotransmitters and neurochemicals necessary for normal mental performance .

3-Strengthening the immune system in general by increasing the production of antibodies and immunoglobulins, which are essentially proteins of the immune system .

4-Participating in some biochemical reactions to remove excess toxins from the cells, such as ammonia, which has been found to be very harmful to the nervous system, brain, and liver .

-Glutamic acid (Glu): Also known as glutamate, it is an alpha amino acid and an excitatory neurotransmitter known in the central nervous system. Glutamic acid is similar to aspartic acid, but the difference between them lies in the side chain, where the side chain of glutamic acid has an additional methylene group ($-CH_2$)compared to the side chain of aspartic acid.

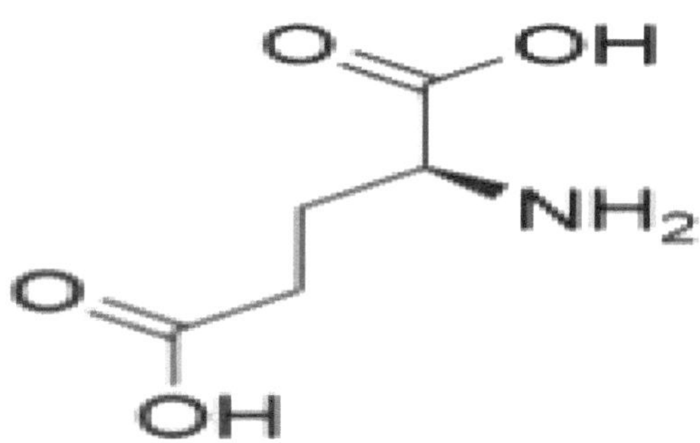

Structures of Glutamic acid

Glutamic acid is a negatively charged amino acid due to the presence of a negative charge in its side chain (carboxyl group). Glutamic acid is naturally synthesized in the body and is also obtained from dietary sources. Glutamic acid helps in the biosynthesis of proteins in the human body, It also helps in the metabolic process, improving brain performance, and heart function. It also participates in many biochemical pathways such as glucose formation and ammonia detoxification. Additionally, glutamic acid supports the normal function of the prostate.

-Glutamine (Gln): It is one of the most common amino acids in the body. It is an alpha-type amino acid containing five carbon atoms, with a molecular weight of 146.15 kilodaltons. Its elemental composition consists of carbon, oxygen, nitrogen, and hydrogen. Glutamine is classified as a non-essential amino acid nutritionally, while physiologically, in terms of its acidity, it is considered a neutral amino acid.

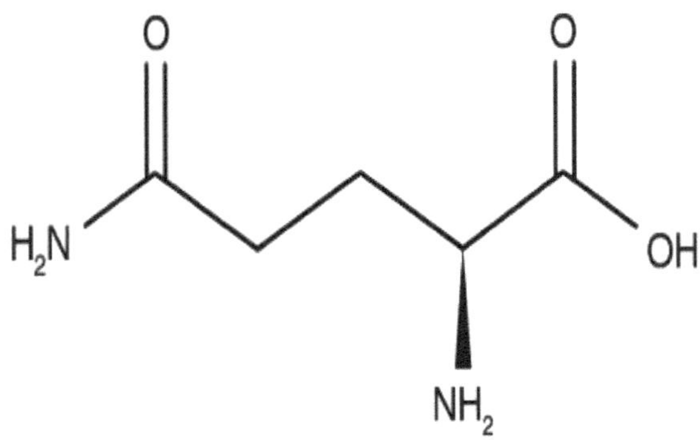

Structures of Glutamine

Glutamine acts as a nitrogen carrier and ammonia transporter due to its content of two amino groups: the alpha amino group and the side chain amide group, which are easily hydrolyzed.

The amount of glutamine in healthy people weighing around 70 kg is approximately 70 to 80 grams of glutamine distributed throughout the body. The concentration of glutamine in blood plasma taken from a person's blood samples after 12 hours of fasting ranges between 500 to

800 micromoles per liter, representing about 20% of the total free amino acids in the blood.

The concentration of glutamine in tissues such as the liver and skeletal muscles is higher than in plasma, where it represents about 40% to 60% of the total amino acids.

Glutamine is considered the most abundant amino acid in the body due to its concentration in both plasma and tissues, where it is higher than any other amino acid by 10 to 100 times. The concentration and presence of glutamine depend on the balance between glutamine synthesis and its secretion, as well as its absorption by human organs and tissues.

The lungs, liver, brain, skeletal muscles, and fat are considered tissues that have distinct activity in the formation of glutamine.

Glutamine is used to provide energy to the digestive system, and it also enhances alcohol metabolism, which helps protect the liver. Additionally, glutamine contributes to protecting the stomach and the digestive system.

-Arginine (Arg): It is a semi-essential amino acid that plays a central role in various aspects of body functions and metabolic processes. It is classified as semi-essential because the body can produce it from glutamine, glutamate, and proline. Dietary intake is considered the main determinant of plasma arginine levels, as the rate of endogenous arginine synthesis does not increase to compensate for deficiency or insufficient supply. Each arginine molecule contains four nitrogen atoms, making it the most abundant nitrogen carrier in humans and animals.

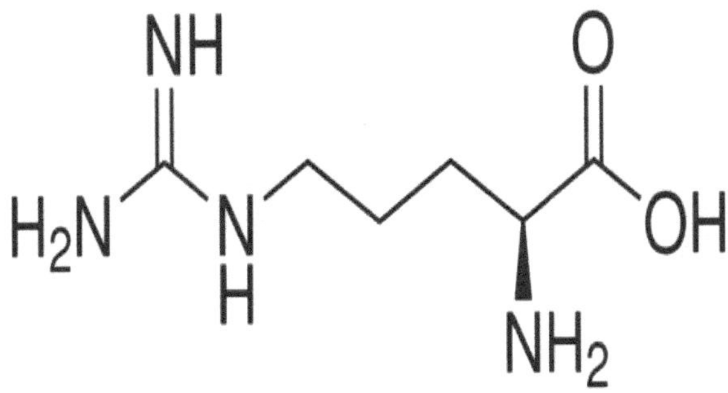

Structures of Arginine

Arginine is essential for the detoxification of ammonia toxins because it plays an important role in nitrogen metabolism as an intermediate in the urea cycle, although arginine is not a primary carrier of nitrogen between organs.

Most of the endogenous arginine in adults is derived from citrulline , which is a byproduct of glutamine metabolism in the intestines or liver, where citrulline is released into the bloodstream and primarily absorbed by the kidneys to convert it into arginine.

Arginine is considered the biological precursor of nitric oxide (NO), which is of great importance in protecting against oxidative damage, maintaining normal blood pressure, heart muscle function, programmed cell death, and the inflammatory response. Arginine is also an essential component of the antidiuretic hormone (ADH) (vasopressin).

-Proline (Pro): It is the only protein amino acid that is a secondary amine, where the nitrogen atom is bonded to both the alpha carbon atom and a chain of three carbon atoms that together form a pyrrolidine ring. This classifies it as an aliphatic amino acid. Proline is distinguished as the only amino acid whose side chain is linked to its main structure twice: once to the central carbon atom and once again to the nitrogen atom in the amino group. This double attachment forms a unique ring structure with secondary nitrogen. Proline is considered a hydrophobic amino acid.

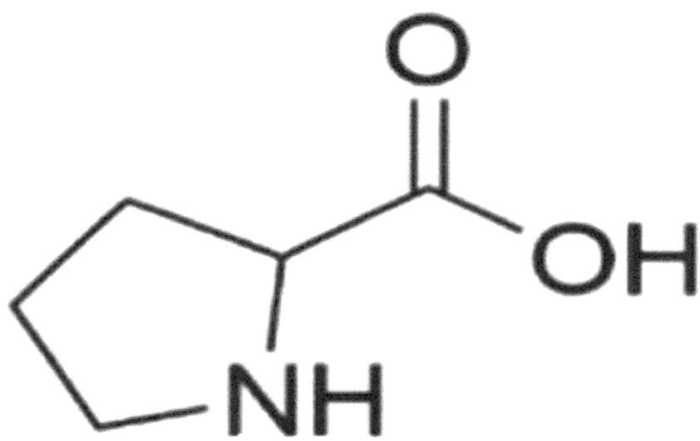

Structures of Proline

The human body synthesizes proline from a non-essential amino acid called (L-glutamate), thus it is a non-essential amino acid in humans.
The German chemist Richard Willstätter isolated proline for the first time in 1900, obtaining proline during his study of the compound (N-methylproline) and synthesizing it by reacting the sodium salt of diethyl malonate with 1,3-dibromopropanol.
Proline is found in most proteins and is also responsible for skin strength. It is also involved in the structure of collagen, the main component of connective tissues, cartilage, ligaments, bones and teeth. Proline has major uses in maintaining blood pressure, preventing arteriosclerosis, and repairing tissues.

-Serine (Ser): It is a non-essential alpha amino acid used in the biosynthesis of proteins, containing an alpha amino group, a carboxyl group, and a side chain consisting of a hydroxymethyl group, which makes it a polar amino acid.

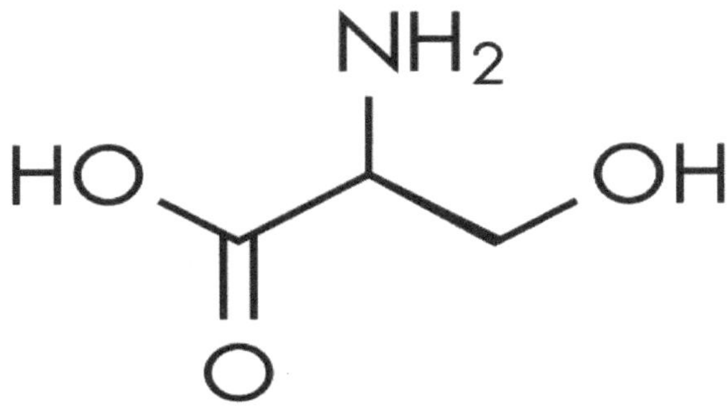

Structures of Serine

Serine can be synthesized in the human body under normal physiological conditions, making it a non-essential amino acid. It is also obtained from food, where it is found in nuts, chickpeas, lentils, meats, fish, and eggs.
In 1865, serine was first discovered in nature, specifically in silk protein, by the German chemist Emil Cramer. Its name was derived from the Latin word "sericum" which means silk. In 1902, the structure of serine was determined.
Serine plays an important role in the metabolism of ketones and glycine, and serine is involved in the production of glycine and other molecules in nerve cells.
There are studies that indicate that serine improves alertness, attention, verbal fluency, and also improves memory in older adults with cognitive problems.

-Glycine (Gly): also called (aminoacetic acid, aminoethanoic acid), is a non-essential aliphatic amino acid consisting of a single hydrogen atom in its side chain. It is the simplest stable amino acid with multiple functions and effects, and it is the smallest amino acid with a molecular weight of 75.067 g/mol.

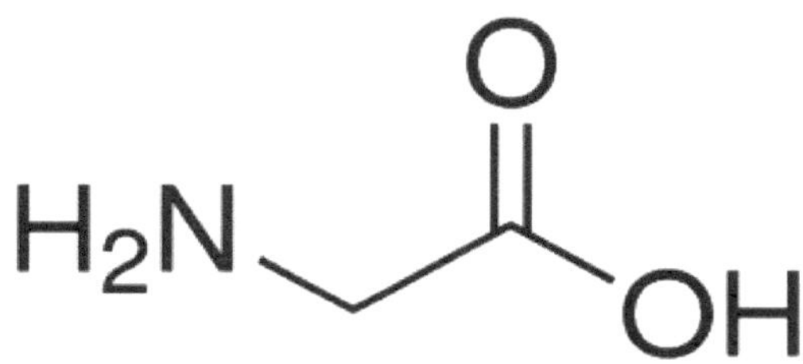

Structures of Glycine

Glycine is an important component of many proteins and plays a significant role in the synthesis of various biomolecules, including creatine and purine nucleotides. It plays an important role in peptide and protein chains and is also considered an inhibitory neurotransmitter. Additionally, it is used to help build muscle tissues, convert glucose into energy, and is essential for maintaining the health of the digestive system and the central nervous system.

In 1820, the French chemist Henri Braconnot discovered glycine through the acid hydrolysis of gelatin and named it gelatin sugar.

The name glycine is derived from the Greek word (glykys), which means sweet. Glycine is abundant in plasma, representing 11.5% of the total amino acids, and constitutes 80% of the protein.

-Cysteine (Cys): A non-essential amino acid containing sulfur. Cysteine in proteins can form disulfide bonds essential for protein folding. Cysteine is obtained from the diet and is also produced internally through the sulfur conversion pathway. Cysteine is an important source of sulfur in human metabolism. It has various roles in protein functions and oxidative metabolism. Cysteine is the main protein in nails, skin, and hair, found in beta-keratin, and is also important in collagen formation.

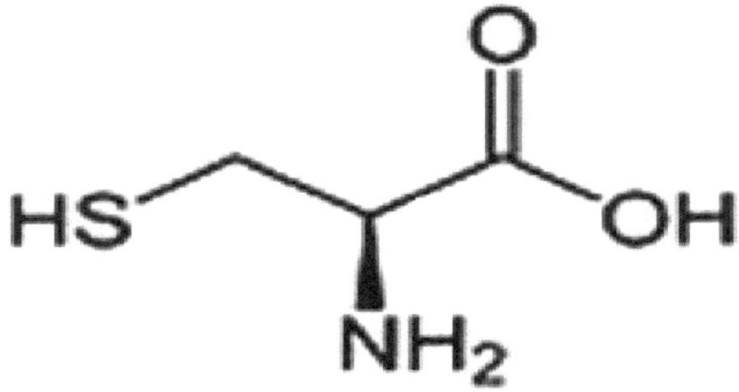

Structures of Cysteine

The thiol group plays a major role in many vital functions of cysteine, such as enabling the formation of disulfide bonds which are essential for protein structure, providing resistance to proteolytic degradation, facilitating the catalytic activity of enzymes, and stabilizing extracellular proteins. Cysteine is also used in baking to break down gluten in flour, which reduces its viscosity and makes the dough easier to knead. Additionally, cysteine and its derivatives, such as acetyl cysteine, are used in expectorant medications (cough syrups), where acetyl cysteine helps break down mucous proteins in the airways, thereby relieving mucus.

Cysteine also plays an important role in the biosynthesis of glutathione and trypanothione. Cysteine is the precursor to the biosynthesis of many essential molecules, including coenzyme A, coenzyme factors, and widely distributed iron and sulfur groups. These molecules are essential for many functions such as regulating oxidation-reduction processes, electron transport, nitrogen fixation, and sensing regulatory processes. Dysregulation of cysteine metabolism occurs in many neurodegenerative diseases (Huntington's disease) and during aging. One of the metabolic products of cysteine is hydrogen sulfide (H2S), which regulates many cellular processes.

In 1810, cysteine was discovered, and in 1899 it was recognized as a component of proteins when it was isolated from animal horns.

-Tyrosine (Tyr): A non-essential amino acid produced from phenylalanine, which means that in addition to being obtained from the diet, it is also synthesized within the body. This amino acid has a polar aromatic side chain but is uncharged (hydrogenated phenylalanine). Tyrosine is found in casein, a protein present in milk and other dairy products such as cheese. Tyrosine is also found in aged foods, including certain types of meat and red wine. Tyrosine contributes to the synthesis of thyroid hormones, melanin, and catecholamines. Tyrosine is the precursor for three important catecholamine neurotransmitters (dopamine, epinephrine, norepinephrine) that affect concentration, mood, alertness, and memory. Tyrosine is essential for the functioning of organs responsible for hormone production and regulation, including the adrenal gland, pituitary gland, and thyroid gland. It also participates in melanin production.

Abnormal levels of tyrosine are closely associated with several diseases such as depression, dementia, Parkinson's disease, vitiligo, and obsessive-compulsive disorder.

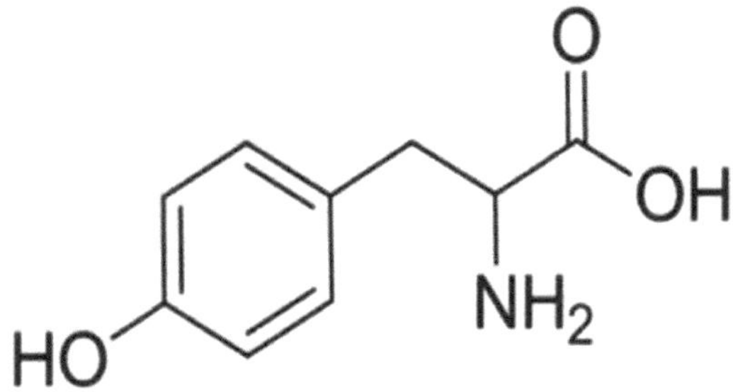

Structures of Tyrosine

Tyrosine is often added to protein supplements to treat a genetic disorder called phenylketonuria (PKU), a genetic disorder that increases phenylalanine levels in the blood. It is also used to treat cocaine addiction, alcohol use disorder, and to improve memory and cognitive skills.

In 1846, the German chemist Justus von Liebig first discovered tyrosine in the casein protein of cheese. The word "tyrosine" is derived from the Greek word "tyros," meaning cheese.

The recommended daily intake of tyrosine is 25 mg/kg of body weight. When excessive amounts of tyrosine are consumed, it is broken down through phosphorylation, sulfation, oxidation, and other metabolic processes.

Classification of amino acids based on the structure of the side group

Amino acids are classified into three groups based on the structure of the side group, where amino acids are divided into three main categories depending on the number of amino and carboxyl groups they contain. These categories are: neutral, basic, and acidic.

1)Neutral Amino Acids: This type of amino acids contains one carboxyl group and one amino group, thus they are neutral, but they differ in the carbon chain attached to the alpha carbon atom (-α) adjacent to the carboxyl group. Accordingly, they are classified into:

A)Aliphatic amino acids: These include glycine, alanine, valine, leucine, and isoleucine.

B)Sulpho amino acids: These contain sulfur, such as cysteine and methionine.

C)Aromatic amino acids: These contain an aromatic benzene ring, such as phenylalanine, tyrosine, and tryptophan (indole ring).

D)Heterocyclic amino acids: These contain a heterocyclic ring, such as the amino acid proline.

2)Basic amino acids: This type of amino acids contains one carboxyl group and two or more amino groups, such as lysine, arginine, and histidine (imidazole ring).

3)Acidic amino acids: These amino acids contain two acidic carboxyl groups (dicarboxylic) and one amino group (one basic group), such as glutamic and aspartic acids.

Protein structure

Proteins are made up of polypeptide chains and have a primary structure, a secondary structure, a tertiary structure, and a quaternary structure, resulting in a complex three-dimensional shape, where proteins are linked to each other by hydrogen bonds, sulphur-sulfur bonds, and weak intermolecular forces.

Primary structure of proteins

The primary structure of a protein is the sequence of amino acids in a polypeptide chain, and the bonds present in this structure are strong peptide bonds, meaning strong covalent bonds. To break these strong bonds within the primary structure of the protein, high temperature and a high concentration of hydrochloric acid are required. The following figure illustrates the primary structure of the protein (the first three letters of each amino acid in the primary structure are represented).

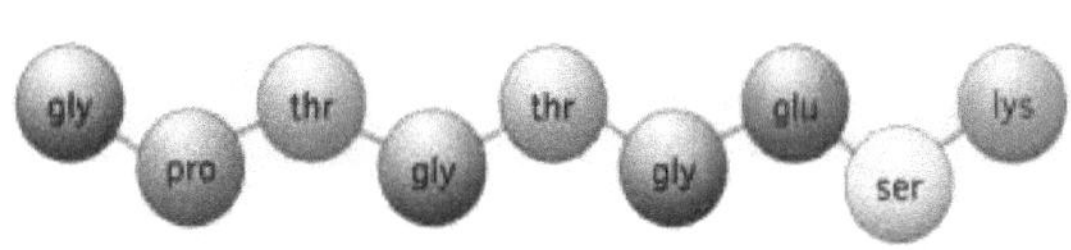

Primary protein structure

The secondary structure of proteins

The secondary structure of a protein is either an alpha helix or folded beta sheets, as shown in the diagram.

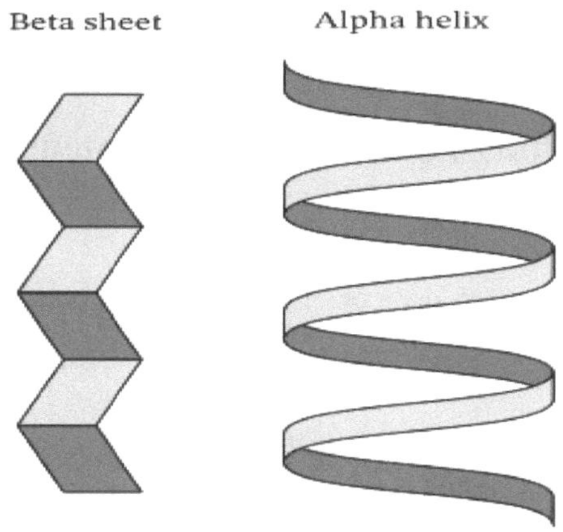

Secondary protein structure

The secondary structure of the protein involves hydrogen bonds between amino acids, where the patterns of hydrogen bonds differ among the various secondary structures, which are:

1)The alpha helix, where hydrogen bonds form between the (C=O) group of one amino acid and the (N-H) region of the amino acid that precedes it by four positions, as illustrated in the diagram.

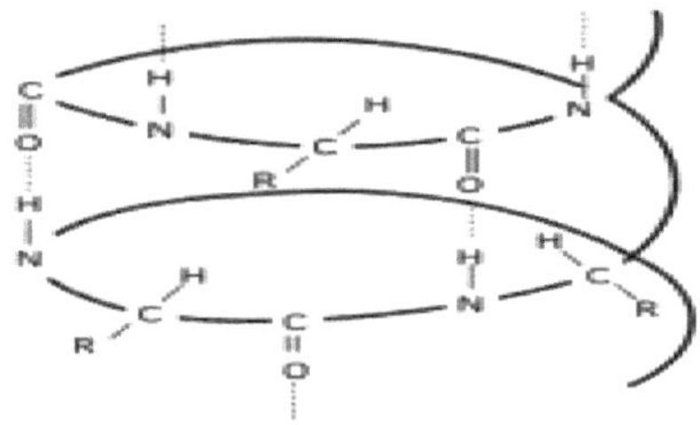

The alpha-helix secondary protein shape

2)The folded beta sheet also consists of hydrogen bonds between (C=O) of one peptide bond and (N-H) groups of another peptide bond, at a later stage of the chain.

The protein may contain several alpha helical regions and other folded beta sheets. Changes in pH or mild heating can alter the secondary structure because hydrogen bonds are much weaker than covalent bonds.

Tertiary Structure of Proteins

The tertiary structure is a three-dimensional structure formed by the folding of alpha helices and/or folded beta sheets. The tertiary structure is held together by:

-Ionic interactions: electrostatic attraction forces between (R) groups with opposite charges, as shown in the following figure.

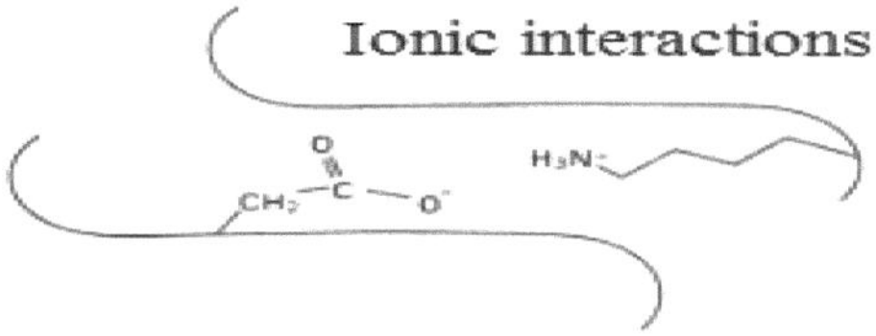

Ionic interactions

-Hydrogen bonding: It forms between (R) groups that allow it, as shown in the following figure.

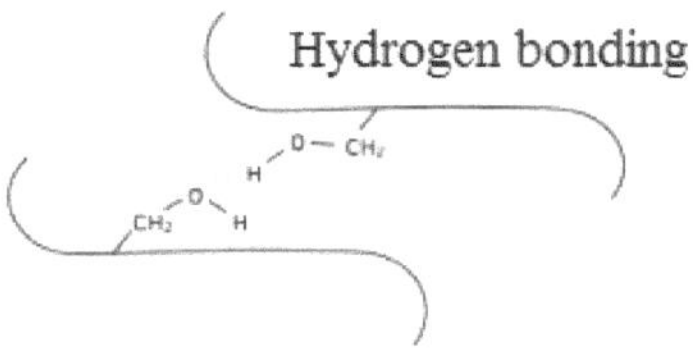

Hydrogen bonding

-Disulfide bonds (sulfur-sulfur bonds): They arise between sulfur atoms within the (R) groups of different amino acids. These covalent (S-S) bonds are strong and help stabilize the tertiary structure of the protein. For example, cysteine is an amino acid containing the group ($-CH_2$ SH), and sulfur-sulfur bonds can form between two of the following:

$$-CH_2SH + HSCH_2- + [O] \rightarrow -CH_2S\text{-}SCH_2 + H_2O$$

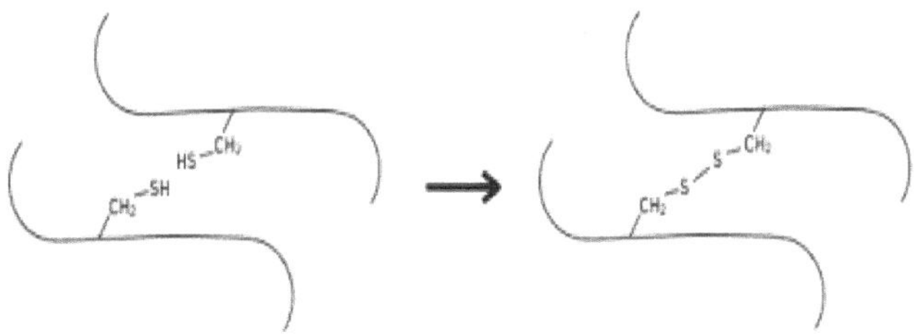

Disulfide bonds

The tertiary structure determines the shape of the protein, and this is what is responsible for how proteins can function in their biological roles, for example, enzymes.

Quaternary structure of proteins

The fourth level of protein structure classification is called the quaternary structure of the protein, which is the arrangement of multiple protein chains or subunits linked by hydrogen bonds and van der Waals forces. Each subunit has its primary, secondary, and tertiary structures to form a functional protein. Any change in this arrangement can significantly affect the biological activity of the protein; therefore, the subunits in the quaternary structure must be precisely arranged for the entire protein to function correctly. The quaternary structure of the protein is formed by the union of small protein units (subunits) to perform a specific function. An example of proteins with a quaternary structure is hemoglobin, which consists of two pairs of different proteins, and each protein binds to a heme molecule (heme is a compound composed of iron(Fe^{2+}) located at the center of a large heterocyclic organic ring called porphyrin, which consists of four pyrrole groups linked together by methine bridges). Each subunit of hemoglobin contains histidine residues that form a covalent bond with the heme, as shown in the following figure.

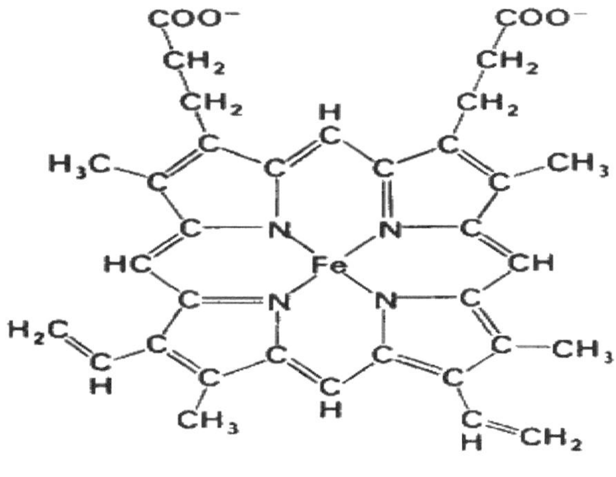

Heme molecule(Porphorin ring with iron atom ligand bound inside)

The two identical alpha chains (α) and the two identical beta chains (β) are arranged in a tetrahedral structure in a three-dimensional configuration, as shown in the figure.

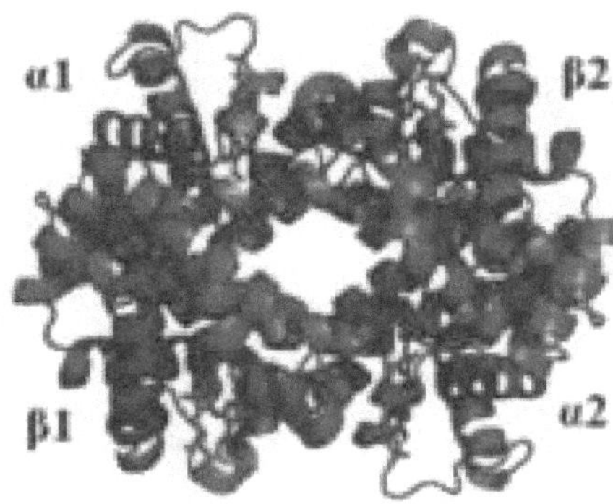

Structure of human hemoglobin

These subunits are linked to each other through hydrophobic interactions, hydrogen bonds, and salt bridges. The four subunits of hemoglobin do not act independently; when one heme molecule binds to oxygen, the conformation of the surrounding polypeptide chain changes slightly. The changes in conformation at one site caused by a change at a spatially separate site in the protein molecule are called allosteric effects. As a result of these allosteric effects, each heme in the other subunits can bind more easily to additional oxygen molecules. With the binding of each oxygen atom, further conformational changes occur in the other protein chains, enhancing their binding ability. Consequently, once oxygenation (the process of adding oxygen to something) occurs in one heme, cooperation happens at all other sites in hemoglobin, allowing it to carry four oxygen molecules.

Protein folding

It is the physical process in which a linear polypeptide chain bends into a three-dimensional structure, thereby becoming a biologically active protein.

Folded proteins are linked through various molecular interactions. During translation, each protein is made as a linear chain of amino acids, or random coils that do not have a stable three-dimensional structure. The amino acids in the chain eventually interact with each other to form a well-defined folded protein. The sequence of amino acids in a protein determines its three-dimensional structure, as folding proteins into their correct native structure is key to their function. Improper folding results in the production of inactive or toxic proteins, leading to dysfunction and causing a number of diseases .

Factors Affecting Protein Folding

Protein folding is a highly sensitive process influenced by many external factors, including magnetic and electric fields, pH level, temperature, chemicals, limited space , and molecular crowding. These factors affect the proteins' ability to fold into their correct functional shapes. Some cells contain heat shock proteins or chaperones that protect the proteins in the cell from thermal degradation. Chaperones help proteins fold and remain folded at extreme temperatures, and they also help misfolded proteins to disassemble and refold correctly.

Chapter Two

Classification of Proteins Based on Structural Shape

Classification of Proteins Based on Structural Shape

Proteins are classified according to the shape of the protein molecule into two categories: globular proteins and fibrous proteins.

Globular Proteins

These are a type of protein that have a spherical and compact shape, with their active surface comprising about 25% of the total protein. These proteins sometimes participate in essential metabolic processes and also have various dynamic functions.

Globular proteins are more sensitive to changes in pH and temperature than their fibrous counterparts. Globular proteins include enzymes, nutrient proteins, cytochrome c, myosin in muscles, blood proteins (serum albumin) , myoglobin, hemoglobin, immunoglobulins (antibodies) , hormones, and glycoproteins.

Globular proteins are large molecular compounds but are not typical polymers. Globular proteins are more complex in their structure than fibrous proteins. These proteins usually dissolve in water or in aqueous media containing alcohol, bases, acids, or salts and disperse easily .

Globular proteins perform a wide range of vital metabolic roles in the human body, including:

1-Enzymes, which catalyze organic reactions that occur in the organism under moderate conditions and with high precision. Examples of these digestive enzymes include amylase, pepsin, and lipase, which break down starch, protein, and fats, respectively.

2-Due to the solubility of globular proteins, they function as messenger proteins or what is known as hormones, transmitting messages to regulate vital processes. An example of this is insulin, which regulates blood sugar levels.

3-Globular proteins act as carriers for molecules because of their soluble nature, allowing them to cross cell membranes. An example of this is hemoglobin, which carries oxygen.

4-Globular proteins have regulatory roles, unlike fibrous proteins.

5-Structural proteins, which form part of the cytoskeleton such as actin and tubulin, are globular and soluble as monomers but polymerize to form long, rigid fibers.

6-Globular proteins serve as reservoirs for amino acids.

Fibrous proteins

Fibrous proteins are long chains composed of repeated sequences of amino acids forming long polypeptide chains. Fibrous proteins are characterized by mechanical strength and insolubility in water because the hydrophobic parts of the polypeptide chains are not folded away from the external environment. They also play a structural role in nature and consist of protein chains arranged in parallel. They include proteins that can swell in water but are insoluble, such as collagen, keratin, and elastin, which are main components of connective tissue, hair, and skin.

In fibrous proteins, a large number of molecules of the same type bind together to form sheets or bundles, where the molecules initially connect through the formation of weak non-covalent bonds, but later these develop into strong covalent bonds, leading to the transformation of the structure into a solid, insoluble mass that is often very strong.

Sometimes structural proteins are fibrous proteins, and because fibrous proteins have stable structures, they are not suitable to function as metabolic proteins. Instead, they work well as structural proteins that support and protect tissues. Examples of fibrous proteins include collagen, keratin, fibrin, and myosin.

Collagen

Collagen is a strong, fibrous, and inextensible protein composed of three intertwined polypeptide chains, where the three collagen chains (alpha chains) twist around each other in a superhelical arrangement to form a rigid rod. The three strands are bonded together by hydrogen bonds involving hydroxylysine and hydroxyproline residues, as shown in the figure.

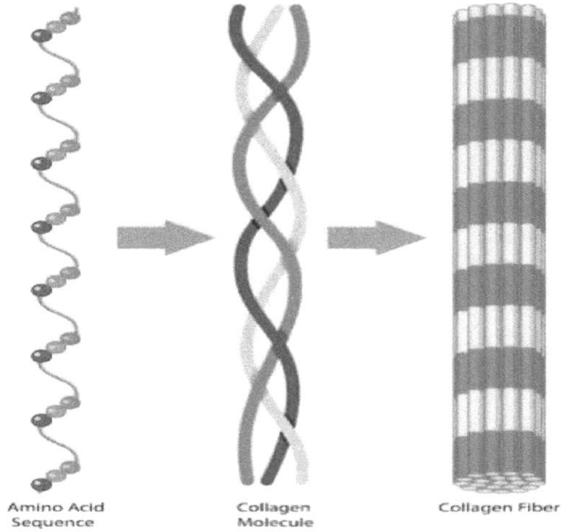

Collagen structure and formation process

Collagen is a major component of connective tissue in animals, including cartilage, tendons, bones, teeth, skin, and blood vessels. Collagen is considered the most abundant protein in vertebrates.

Collagen is the most abundant protein in mammals, is found throughout the body. Collagen protein constitutes a large proportion, up to 25%, of the total proteins in mammals and some other living organisms. Collagen protein provides structural support to the extracellular space of connective tissues. Due to its hardness and resistance to stretching, collagen is considered the ideal matrix for skin, tendons, bones, and ligaments.

Collagen can be divided into several other groups according to the types of structures it forms. Twenty-eight different types of collagen have been discovered, and types I to V are considered the most common, with type I making up more than 90% of the collagen in the human body. These five types are:

1- **Type I:** includes skin, blood vessels, bones, and tendons, and is the main component of the organic part of bones.
2- **Type II:** represents the primary collagen component of cartilage.
3- **Type III:** this type of collagen is found in reticular fibers, muscles, and arteries.
4- **Type IV:** is located in the epithelial layer of the basement membrane of the cell.
5- **Type V:** is one of the main components of cell surfaces, the placenta, and hair.

Keratin

Keratin is a fibrous structural protein found in epithelial cells that line the internal and external surfaces of the body. It is also found in the cells lining the organs, glands, and other parts of the body. Keratin helps in forming the tissues of hair, nails, and the outer layer of the skin.
Keratin plays an important structural and protective role, and some types of keratin have been found to regulate key cellular activities such as cell growth and protein synthesis.

Keratin is completely insoluble in water due to its high content of hydrophobic amino acids, and it is not affected by proteolytic enzymes. It contains a high sulfur content, as keratin has a large amount of cysteine.

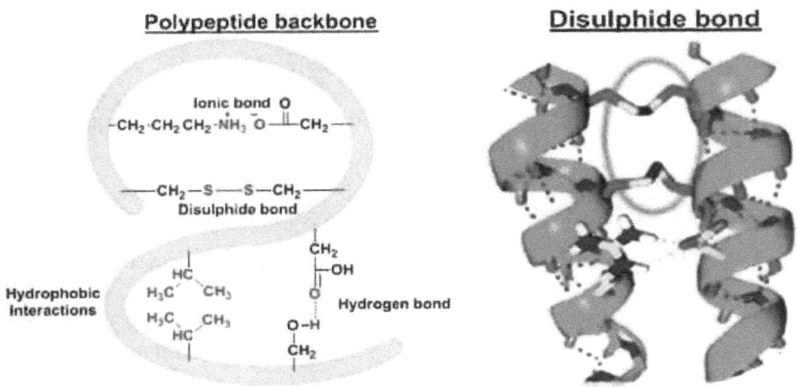

Polypeptide backbone and 3D structure of keratin protein

Keratin is classified based on its secondary structure into two types: alpha-keratin and beta-keratin:

1-**Alpha-keratin:** It is found in the hair, skin, and nails of mammals, characterized mostly as fibrous, with a structure resembling a helical coil (spiral). This form of keratin in the human body is divided into two types, the first and the second.

2-**Beta-keratin:** It consists of sheets of polypeptide chains arranged in parallel, which gives this form of keratin a solid and durable structure. It is found in bird feathers, reptiles, their scales, claws, and beaks.

There are 54 types of keratin in the human body, divided into two types:

-**The first type:** contains 28 types, divided into 11 types of hair keratin and 17 types of skin cell keratin. Most types of keratin of this type (cellular keratins) consist of lightweight acidic proteins and have many functions, including helping to protect cells from internal forces in the human body.

-The second type: contains 26 types of keratin in the human body, with six types of hair keratin and 20 types of skin cell keratin, consisting of high molecular weight neutral basic proteins.
Some types of keratin may be present in higher than normal amounts in patients with various types of epithelial cell cancers, including breast, lung, bladder, colon, rectal, neck, and head cancers.

The structure of the central rod domain of a typical alpha-keratin consists of pairs of right-handed helices coiling around each other to form a left-handed supercoil, as shown in the figure.

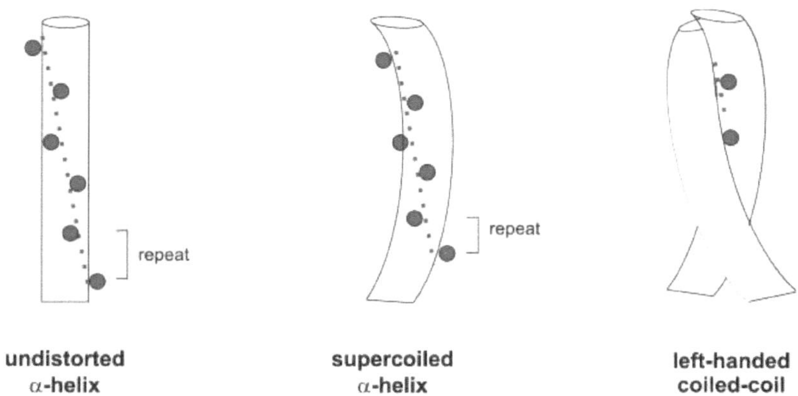

The central rod domain of a typical α-keratin

Fibrin

Fibrin is a fibrous, non-globular protein also known as Factor I (a) . It is solid and arranged in long fibrous chains and participates in blood clotting. Fibrin is produced from another soluble protein produced by the liver called fibrinogen, which is found in blood plasma. When tissues are damaged, leading to bleeding, fibrinogen is converted into fibrin at the wound site by the action of the clotting enzyme thrombin. Then, fibrin molecules unite to form long fibrin threads that intertwine with platelets, resulting in the formation of a spongy mass that gradually hardens and contracts to become a blood clot. This hardening process is stabilized by a substance known as fibrin-stabilizing factor, or Factor XIII.

Liver disease or dysfunction can lead to a decrease in fibrinogen production or the production of abnormal fibrinogen molecules with low activity (dysfibrinogenemia), resulting in impaired fibrin production.

Fibrin-like substances may sometimes be present in higher than normal amounts in the blood and urine of patients with certain types of cancer. Measuring the amount of these substances may help verify the effectiveness of cancer treatment or whether the cancer has worsened. In 1666, the Italian biologist and physician Marcello Malpighi discovered fibrin.

Myosin

Myosin is a family of motor proteins and a fibrous muscle protein capable of hydrolyzing adenosine triphosphate (ATP). It interacts with actin during muscle contraction to form actomyosin, the actin–myosin complex that is inherently contractile within the cytoskeleton.

Each member of the myosin family has unique functional and structural characteristics, such as its step size, which determines its ability to engage in the movement of actin filaments.

There are at least 35 distinct classes of myosins, responsible for a wide range of cellular processes that require force generation and movement, including cell motility, cytokinesis, organelle transport, and muscle contraction. All myosins share specific structural features.

Type II myosin is found in skeletal muscle cells. Type II myosin forms only thick filaments that are arranged within a structure of thin actin filaments (in addition to many other proteins), where these filaments form complex fibrous structures called sarcomeres.

A single sarcomere consists of repeating units of thick (myosin) and thin (actin) filaments. During muscle contraction, the opposing orientation of myosin heads pulls adjacent actin filaments toward one another, causing the filaments to slide. The sarcomere also contains transverse tubules (T-tubules) that transmit electrical signals and help regulate the release of calcium ions (Ca^{2+}), as well as mitochondria that provide the energy required for contraction. All these components are enclosed within a specialized cytoplasm known as the sarcoplasm.

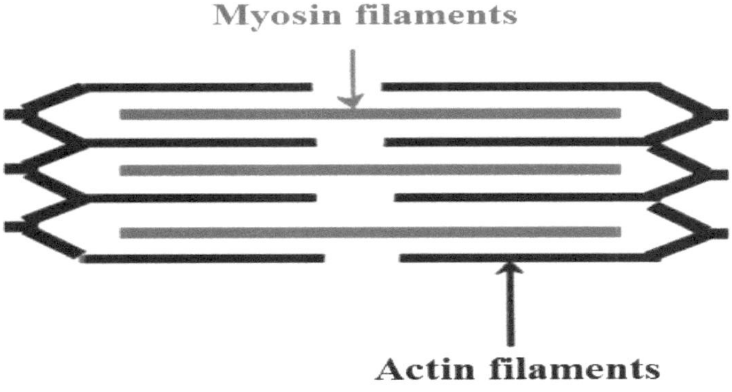

An illustration of a sarcomere

The length of a sarcomere in resting muscle is approximately 2 micrometers, and during muscle contraction, this length can shorten by up to 70%. Muscle contraction is regulated by calcium ions and the troponin-tropomyosin regulatory system.

Although the actin units continue to rotate at both ends of the thin filament, this exchange is relatively slow, making the actin filaments in sarcomeres relatively more stable compared to the actin filaments found in other types of cells.

In the case of non-muscle cells, myosin II binds to actin filaments to form contractile structures called stress fibers, which extend along the basal surfaces where the cell adheres to its substrate, as in epithelial cells. Contractile bundles also appear in the adhesion belt, helping to maintain the stability and integrity of epithelial cell sheets.

The formation and contractile activity of the myosin II bundle in non-muscle cells are regulated by phosphorylation.

Myosin was first discovered in 1864 by the German scientist Wilhelm Friedrich Kuhn.

Chapter Three

Classification of Proteins According to Their Chemical Structure

Classification of Proteins According to Their Chemical Structure

Proteins are divided into two main categories based on their behavior when they react with water: simple proteins and conjugated proteins.

Simple Proteins

They consist of one or more polypeptide chains and are also known as homogeneous proteins. A single polypeptide yields only amino acids upon hydrolysis, without any additional non-peptidic components, as they are linked together by peptide bonds. They are classified based on their structure and solubility and may also contain non-protein prosthetic groups, as is the case with enzymes.

Simple proteins play an important role in vital functions, including structural support, transport mechanisms within living organisms, and enzymatic activity.

They can be divided into different subtypes such as globular and fibrous proteins. Examples of globular proteins include albumins, histones, globulins, hemoglobin, and insulin, while fibrous proteins include collagen and keratin.

Simple proteins can also be classified according to their solubility, and although this type of classification may not be optimal for this complex group of compounds, they are classified based on their solubility as follows:

1)Albumins: soluble in water, dilute salts, bases, and acids.

2)Histones: basic proteins that are soluble in water but do not dissolve in dilute ammonia solution.

3)Protamines: similar to histones but do not coagulate with heat like other soluble proteins.

4)Globulins: They are insoluble in pure water but dissolve in acids, bases, and diluted salts.

5)Glutelins: They dissolve in diluted acids or bases but are insoluble in neutral solvents, including salt solutions.

6)Prolamins: They are soluble in alcohol and this type of protein contains a high proportion of proline.

7)Insoluble proteins: They are insoluble in most solvents.

Conjugated Proteins

Also called heterogeneous proteins, these are proteins that, in addition to amino components, contain non-amino components which may include carbohydrate residues(glycoproteins and proteoglycans), lipids (lipoproteins), phosphate groups (phosphoproteins), metal ions such as (Fe^{2+} or Ca^{2+}), (metalloproteins), or nucleic acids (nucleoproteins such as histones).

Conjugated proteins are classified according to the chemical nature of the attached group as follows:

Glycoproteins - Lipoproteins - Phosphoproteins - Metalloproteins - Nucleoproteins.

Glycoproteins

Glycoproteins are conjugated proteins usually found on the surface of cells where these proteins covalently bind with carbohydrates through glycosidic bonds. Their short carbohydrate chains consist of sugar residues extending outside the cell, where the carbohydrate domain is essential for cell adhesion, cell recognition, intercellular communication, and signal transduction.

These proteins play a vital role in many physiological functions, including the immune system. Many viruses contain glycoproteins that help them enter the body's cells. Additionally, these proteins can be important targets for the prevention or treatment of a wide range of diseases, such as respiratory, kidney, cardiovascular diseases, as well as tumors, diabetes, aging, and other conditions.

Glycoproteins vary greatly in the composition, number, and size of their carbohydrate side chains. Glycoproteins are abundantly present throughout nature. The main sources of glycoproteins are algae, microorganisms, plants, and animals. Among the most prominent glycoproteins are glycoprotein and proteoglycan.

1-Glycoprotein: These are conjugated proteins that contain one or more short, irregular side chains of heteropolysaccharides covalently linked to a polypeptide chain. Glycoproteins in some aspects resemble proteoglycan proteins but differ from them in the shortness of their oligosaccharide chains. Glycoproteins are proteins bound to oligosaccharide chains, not to regularly repeating disaccharides. These proteins are found in saliva and are responsible for its surface activity and properties. They are also likely present in the ground substance of many connective tissues but in small amounts.

2-Proteoglycan: It is a type of glycoprotein with a high molecular weight and is specifically found in connective tissue (the fibrous tissue that supports the body's structure). Proteoglycan forms a large part of this tissue as it fills the spaces between cells. These proteins are composed of glycosaminoglycan chains covalently linked to a protein core.

Proteoglycan is produced in the rough endoplasmic reticulum, then reshaped in the Golgi apparatus, and afterward transported to the extracellular matrix via vesicles.
Proteoglycans can be classified according to the glycosaminoglycan chains attached to the core protein, and their types include chondroitin sulfate, heparin sulfate, heparan sulfate, keratan sulfate, and dermatan sulfate.

Proteoglycans are also classified according to their relative size into small and large. Small molecules include: decorin, fibromodulin, biglycan, and lumican, while large molecules include: aggrecan, which is an important part of cartilage (giving cartilage the property of water binding to form hydrated matrices), and versican, which is found in blood vessels and skin.

Lipoproteins

Lipoproteins are large molecular compounds that are spherical, consisting of a hydrophobic core made up of cholesterol ester molecules and triglycerides, surrounded by an outer layer of proteins called apolipoproteins, phospholipids, and cholesterol molecules that are hydrophilic, oriented towards the aqueous environment, and are bound by non-covalent forces.

Lipoproteins are more complex than glycolipids (which are fats linked to carbohydrates and play an important role in energy supply). The primary function of lipoproteins is to transport and deliver fatty acids, cholesterol, and triglycerides to and from target cells in various organs.

In plasma, lipoproteins can be divided into seven categories based on size, lipid composition, and apolipoprotein. These proteins are [chylomicrons, chylomicrons remnants, very low-density lipoprotein (VLDL), intermediate-density lipoprotein (remnants of very low-density lipoprotein) (IDL), low-density lipoprotein (LDL), high-density lipoprotein (HDL), lipoprotein(a) [(Lp(a)] (which is a type of low-density lipoprotein (LDL)].

1-Chylomicrons: They are lipoproteins that are the largest in size and the least dense, containing a high percentage of triglycerides. Chylomicrons are produced in the intestinal cells from dietary fats such as fatty acids and cholesterol. Chylomicrons consist of a central lipid core mainly composed of triglycerides, in addition to containing cholesterol esters and phospholipids like other lipoproteins.

Chylomicrons transport fats and cholesterol, where they enter the lymphatic system and then are released into the bloodstream. After that, they are broken down by enzymes such as lipoprotein lipase (LPL) in the blood, which separates the fatty acids to be used or stored. Then, the Chylomicrons rich in cholesterol are transported to the liver, where they are absorbed by liver cells and the cholesterol within them is metabolized. Afterward, the liver can use this cholesterol to form other lipoproteins.

2-Chylomicron remnants: These are lipoproteins formed as a result of the removal of triglycerides from chylomicrons by muscles and lipoprotein lipase (LPL) in adipose tissues, leading to smaller particles. These particles are rich in cholesterol and are a cause of atherosclerosis. Chylomicron remnants are removed from the bloodstream exclusively by the liver, possibly because these large complexes can penetrate the unique sinusoidal vascular spaces.

3-Very Low-Density Lipoprotein (VLDL): It is the only lipoprotein that contains the apolipoprotein B (ApoB) type that is secreted by the liver. Very low-density lipoprotein is composed of cholesterol, triglycerides, and the apolipoprotein B-100 core. The lipid content in very low-density lipoprotein is about 90%, while proteins make up about 10%. This ratio gives the protein a low density ranging between 0.96 and 1.006 g/ml. Triglycerides constitute about 70% of the mass of very low-density lipoprotein, while cholesterol esters and fatty acids make up the remaining portion.

The primary function of very low-density lipoprotein is to transport cholesterol, other fats, and proteins to peripheral cells to carry out essential biological activities. It is also known that very low-density lipoprotein contributes to the development of cardiovascular diseases and atherosclerosis. In addition, this protein plays an important role in normal biological processes.

Very low-density lipoprotein (VLDL) large particles show a strong association not only with atherosclerosis but also with insulin resistance and the onset of diabetes.

In the liver, triglycerides and cholesterol are combined with apolipoprotein B100 (ApoB100), which affects the amount of fat and the size of the very low-density lipoprotein secreted.

After the secretion of very low-density lipoprotein (VLDL), the lipoprotein lipase enzyme (LPL) located on the capillary endothelium breaks it down, converting the triglycerides in the very low-density lipoprotein into free fatty acids and glycerol, allowing cells to absorb them for use as energy or storage. Very low-density lipoprotein (VLDL) binds to low-density lipoprotein (LDL) receptors, leading to its transformation in the bloodstream into low-density lipoprotein (LDL) remnants. The normal value of very low-density lipoprotein in the human body is less than 30 mg/dL.

4-Intermediate-density lipoprotein (remnants of very low-density lipoprotein) (IDL): It is a type of lipoprotein that forms as an intermediate step (transitional phase) in the normal metabolism of very low-density lipoproteins and is metabolized quickly; therefore, its concentration in plasma is usually very low.

This is done by the breakdown of triglycerides (TG), which are abundant in very low-density lipoproteins, by the enzyme lipoprotein lipase in muscles and adipose tissues. As a result, very low-density lipoproteins are converted into intermediate-density lipoproteins, which are rich in cholesterol and promote atherosclerosis, and therefore are considered harmful. Intermediate-density lipoprotein particles contain the apolipoprotein B100 (Apo B100) and the apolipoprotein E (Apo E).

The relative density of intermediate-density lipoprotein ranges between 1.006 and 1.019. The density of the intermediate-density lipoprotein particle lies between the density of very low-density lipoprotein particles and low-density lipoprotein particles. Typically, the size of the intermediate-density lipoprotein particle ranges between 25 and 35 nanometers.

5-Low-density lipoprotein (LDL): Also known as "bad cholesterol," it is one of the seven main groups of lipoproteins that transport fats molecules throughout the body in the extracellular fluid, making fats available to body cells for endocytosis (the process by which cells absorb the molecules they need) via receptors. Low-density lipoprotein is also associated with the development of atherosclerosis.

This lipoprotein consists of a monolayer of phospholipids where unesterified cholesterol forms the surface membrane, while the fatty acid esters of cholesterol form the hydrophobic core.

The diameter of a single molecule of low-density lipoprotein ranges between approximately 22 and 27.5 nanometers. Each molecule of this protein typically carries between 3000 and 6000 lipid molecules. The normal value for low-density lipoprotein levels in the human body is less than 100 mg/dL.

6-High-density lipoprotein (HDL): also called (good cholesterol), it is the smallest lipoprotein and has a high density, therefore it is called high-density. It is also characterized by having the highest ratio of protein to cholesterol.

High-density lipoprotein usually carries about 25 percent of the total cholesterol present in the bloodstream of a healthy human, while low-density lipoprotein carries most of the remaining amount.

This protein works to transport fat molecules out of the artery walls and carries them to the liver for reuse or disposal. It also helps prevent or reduce atherosclerosis by decreasing the accumulation of macrophages (phagocytes). Additionally, the molecules of this protein possess antioxidant, anticoagulant, anti-inflammatory, and anti-apoptotic properties, which may also contribute to their ability to prevent atherosclerosis. The normal value of high-density lipoprotein protein levels in the human body is above 60 mg/dL.

7-Lipoprotein (a) [Lp(a)]: One of the types of low-density lipoproteins, lipoprotein (a) was discovered in 1963 by the Norwegian scientist Dr. Kåre Berg . It is characterized by the presence of a Lipoprotein A which binds with lipoprotein (B-100). Due to the atherogenic and thrombogenic properties of this lipoprotein, it is considered a fixed and genetically determined risk factor for atherosclerosis, coronary artery disease, thrombosis, aortic stenosis, and stroke.

The normal level of this protein is less than 30 mg/dL, and when its levels in patients are greater than 50 mg/dL, they are more likely to develop cardiovascular diseases due to plaque buildup in the arteries.

Functions of lipoproteins

There are four main functions of lipoproteins:

1)They play a structural role by providing the physical structure for packaging and transporting fats.

2)Lipoproteins located on the outer surface of lipoproteins act as binding agents that interact with specific receptors present on the cell membrane.

3)They transport fats (such as triglycerides and cholesterol) from the intestines to body cells, and from the liver to other tissues.
4)They act as activators or inhibitors of enzymes involved in the metabolism of lipoproteins.

The importance of lipoproteins

Lipoproteins show varying patterns associated with the risk of fatal cardiovascular diseases. High levels of low-density lipoprotein, very low-density lipoprotein, cholesterol, and triglycerides are linked to an increased risk of atherosclerosis and heart diseases.

Smoking raises the level of low-density lipoprotein in the blood while reducing the level of high-density lipoprotein in the blood, which may lead to the accumulation of plaques in the arteries (atherosclerosis). This results in narrowing of the arteries, increasing the risk of heart diseases, heart attacks, and strokes.

The main importance of chylomicrons lies in transporting triglycerides from the intestines to the liver, fat cells, and muscles, while chylomicron remnants transport cholesterol from the intestines.

Low-density lipoprotein (LDL) transports cholesterol to different tissues in the body such as the gonads, adrenal gland, adipose tissues, and muscles.
The main function of very low-density lipoprotein (VLDL) is to transport

triglycerides and cholesterol to the places that need them, thereby helping the body to acquire energy, store it, and regulate blood pressure.

High-density lipoprotein (HDL) is important because of its ability to carry cholesterol from various parts of the body to the liver, where the liver removes cholesterol from the body. It also has antioxidant properties (including protecting low-density lipoprotein from oxidation) as well as anticoagulant and anti-inflammatory properties.

Phosphoproteins

Phosphoproteins are proteins that are usually phosphorylated by kinases, and they can be part of signaling pathways in which one kinase phosphorylates another kinase, or multiple kinases, contributing to the amplification and propagation of the signal within the cell. The first phosphoprotein was discovered in 1883. Some phosphoproteins are characterized by high molecular weights. Phosphoproteins are also known for their ability to resist the activity of protease enzymes (these are enzymes that break down proteins and are produced in animals, plants, fungi, and bacteria).

One of the most phosphorylated proteins in nature is phosvitin, which constitutes 60% of phosphoproteins. It is a lipophilic protein found in egg yolk (usually chicken eggs) and contains 90% of the phosphorus present in egg yolk. It was first isolated from egg yolk in 1949 by Dell Mitcham and Harold Olcott.

Phosvitin has high resistance to the effects of digestive enzymes such as trypsin and protease. Additionally, phosvitin found in egg yolk possesses important properties including the ability to bind calcium and iron, the ability to extract minerals, anti-inflammatory properties, and antioxidant properties.

Metalloproteins

These are proteins in which metal ions are directly bound to some side chains of histidine, cysteine, or other amino acids. The presence of these metal elements contributes to the generation and stimulation of the functions of these proteins, demonstrating their significant importance at the cellular and biological levels. These metals can be transitional, such as zinc, iron, copper, and others.

Metalloproteins represent more than one-third of human proteins, but there is significant variation in physiological functions depending on the metal or metals involved and the tissue context. Their functions include catalysis, oxidation and reduction, cell proliferation, immunity, DNA repair, and the transport of vital elements.

An example of metalloproteins are : hemoglobin, myoglobin, and metal-dependent enzymes.

Hemoglobin

It is a protein found in red blood cells and contains iron (which is responsible for the red color of blood). Hemoglobin is responsible for transporting oxygen from the lungs to the body's tissues and organs to ensure adequate oxygenation of the tissues and returns carbon dioxide to the lungs. It also participates in several other functions such as stimulation, pH regulation, metabolic reprogramming, and maintaining the balance of oxidation and reduction.

The hemoglobin molecule consists of four interconnected protein molecules known as globulin chains, where the natural hemoglobin molecule in adults contains two alpha globulin chains as well as two beta globulin chains, as shown in the figure.

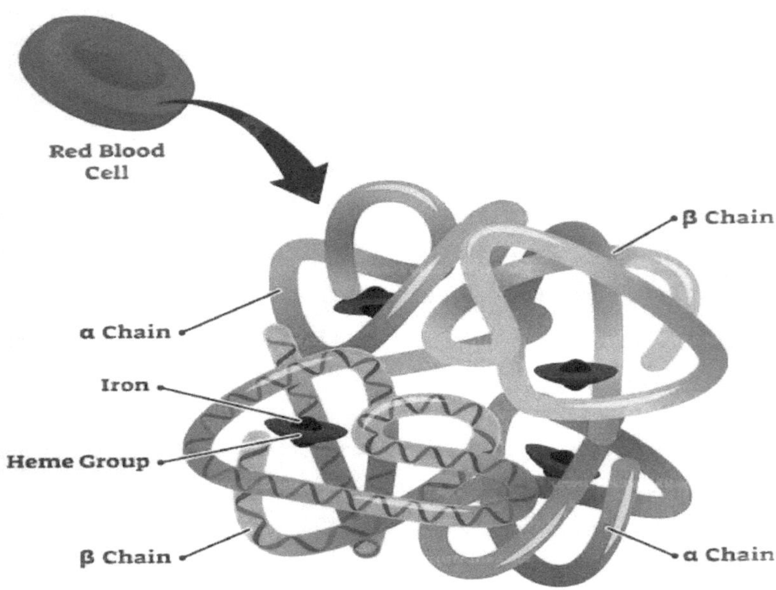

The structure of the hemoglobin molecule

Beta globulin chains are not common in fetuses and infants, but in fetuses and infants, the hemoglobin molecule contains two alpha globulin chains and two gamma globulin chains. As the infant grows, the gamma globulin chains are gradually replaced by beta globulin chains, leading to the formation of the adult hemoglobin structure.

In case of low hemoglobin levels, the patient suffers from anemia, where the amount of hemoglobin in whole blood is measured in grams per deciliter. The normal hemoglobin level for females ranges between 12-16g/dL, for males between 14-18g/dL, and for children it ranges between 11g/dL to 16g/dL.

Myoglobin

It is a protein mainly found in the striated muscles of vertebrates and contains a synthetic heme group capable of reversibly binding oxygen. The body uses it as a protein to store oxygen in the muscles.

Myoglobin is considered one of the members of the globin family, which also includes hemoglobin. Myoglobin is often compared structurally and functionally to hemoglobin. Myoglobin consists of a single polypeptide chain containing one oxygen-binding site, which leads to differences in the oxygen-binding kinetics of the two proteins, whereas hemoglobin contains four polypeptide chains and four oxygen-binding sites.

Myoglobin is mainly found in the striated muscles of vertebrates, while hemoglobin is present in the bloodstream as part of red blood cells.

Myoglobin is also found in much lower concentrations in smooth muscle cells, endothelial cells, and even cancer cells.

The primary function of myoglobin is to supply oxygen to the muscles, which it does by releasing oxygen supplies to the mitochondria that form the respiratory chain, thereby helping muscle cells meet their high energy demands .

Myoglobin also acts as a buffer for oxygen concentrations within cells and as an oxygen reserve in muscles. This supports the fact that marine mammals contain myoglobin levels ranging from 10 to 30 times higher compared to animals that do not hold their breath for long periods. Additionally, myoglobin helps facilitate the diffusion of oxygen, as myoglobin saturation levels decrease at the onset of muscle activity, increasing the oxygen diffusion gradient from the capillaries to the cytoplasm. Myoglobin also possesses enzymatic functions, as it is essential for the breakdown of biologically active nitric oxide into nitrate. The removal of nitric oxide enhances mitochondrial respiration. This enhancement occurs because nitric oxide reversibly inhibits cytochrome oxidase enzyme .

In addition, myoglobin works to remove reactive oxygen species and can do so by interacting with fatty acids, which may be metabolically significant under oxidative conditions and high energy demands. Myoglobin is also considered the cause of the red color in the muscles of most vertebrates.

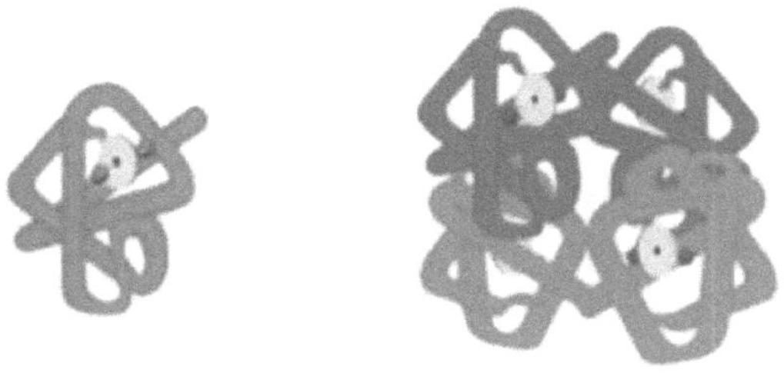

Myoglobin **Hemoglobin**

The structure of hemoglobin and myoglobin

Nuclear proteins

They are proteins that are bound to deoxyribonucleic acid (DNA) or ribonucleic acid (RNA). Nuclear proteins contain phosphoric acid and another basic protein group.

Nuclear proteins play a vital role in living organisms, as they participate in essential processes such as transcription, translation, and replication. They also contribute to regulating metabolic activities within cells and gene expression. They often function in transporting lipids and storing proteins such as ovalbumin (egg white protein).

Chromosomes are a type of nuclear proteins where histone proteins interact with deoxyribonucleic acid, and this is called the nucleosome. Another example of nuclear proteins is ribosomes, where ribosomal proteins interact with ribosomal RNA.

Chapter Four

Classification of Proteins Based on Biological Functions

Classification of proteins based on their functions

Proteins are classified based on their functions as follows:

1-Catalytic proteins such as enzymes .

2-Structural proteins such as collagen and elastin .

3-Contractile proteins such as myosin and actin .

4-Transport proteins such as hemoglobin, myoglobin, albumin, transferrin.

5-Regulatory proteins or hormones such as adrenal cortex hormone, insulin, growth hormone.

1-Catalytic proteins

They are proteins that contribute to stimulating or enhancing cell activity. These proteins also play an important role in various processes such as accelerating chemical reactions, producing immune cells, sending signals, and muscle formation. Some of the most important examples of these proteins are enzymes, proteins that stimulate immune cells, dopamine receptors, and proteins that promote muscle building.

Enzymes

Enzymes are proteins that catalyze chemical reactions inside the living cell with high precision. Enzymes are not consumed during the reaction and are used repeatedly. Enzymes exert their effect on molecules called substrates, the enzyme converts the substrate into molecules called products. Enzymes are known for their ability to catalyze more than 5,000 types of biochemical reactions.

Cells contain many types of enzyme molecules, each specialized for a particular chemical reaction. Since enzymes are proteins, they share the general properties of proteins: they can be denatured or inactivated by heat, strong acids, or bases, which alter their natural structure and function.

Enzymes are composed of one or more long chains of linked amino acids like all other protein molecules. Each enzyme has a distinct shape as it contains a specific sequence of amino acids that leads to its folding and forming a particular shape.

The enzyme requires another non-protein component called a cofactor in order to perform its catalytic activity.

The first person to discover the enzyme was the French chemist Anselme Payen, who in 1833 discovered the enzyme diastase. In 1877, the German physiologist Wilhelm Kühne used the term "enzyme" for the first time, which is derived from the Greek word (ἔνζυμον) meaning "leavened" or "in yeast".

The name of the enzyme is usually derived from its substrate or the chemical reaction it catalyzes, with the suffix (ase) added at the end. An example of this is lactase (which helps digest lactose), alcohol dehydrogenase (important in the metabolism of alcohol), and DNA polymerase (responsible for the process of DNA replication).

The different enzymes that catalyze the same chemical reaction are called enzyme isoenzymes .

The International Union of Biochemistry and Molecular Biology has developed a system for naming enzymes known as the Enzyme Classification Number (EC) in which each enzyme is represented by a four-digit sequence preceded by the symbol "EC". The first digit classifies the enzymes according to their mechanism of action .

The classification is:

EC1-Oxidoreductases: Catalyze oxidation-reduction reactions.

EC2-Transferases: Transfer a functional group (such as a methyl or phosphate group).

EC3-Hydrolases: Catalyze hydrolytic cleavage of various bonds.

EC4-Lyases: Catalyze the breaking of various chemical bonds by means other than hydrolysis and oxidation.

EC5-Isomerases: Catalyze the rearrangement of atoms within a molecule, leading to the conversion of one isomer to another.

EC6-Ligases: Catalyze the bonding of two large molecules by forming a new chemical bond.

EC7-Translocases: Enzymes that facilitate the movement of ions or molecules across a cell membrane, a process essential for cell function.

Immune cell-stimulating proteins

They are essential proteins in the immune system, playing a vital role in regulating and directing the immune response. These proteins act as chemical messengers to exchange information between immune cells, leading to the activation or inhibition of certain cells and triggering various immune responses. Examples of these proteins include cytokines and antibodies.

Cytokines

They are small proteins secreted by cells, usually ranging in size between 5 and 20 kilodaltons, functioning as extracellular signals between cells in immune development and during immune response pathways. They play a key role in mediating interactions and communication between cells. Cytokines are also involved in many biological processes, including tissue repair, initiation and progression of cancer, regulation of immune responses, control of cell proliferation, and the induction of regulated or unregulated cell death.

A large number of cytokines have been identified, which can be roughly classified into five main categories according to their functions: interleukins, tumor necrosis factors, interferons, hematopoietins, and chemokines. Interleukins are regulatory compounds that play a fundamental role in the immune system by either stimulating or inhibiting the activity of immune cells.

Colony-stimulating factors help promote the proliferation and differentiation of white blood cell precursors, while tumor necrosis factors are characterized by anti-tumor activity, playing an important role in stimulating programmed cell death in targeted cells. Interferons play a crucial role in regulating antiviral activity within immune cells. Hematopoietins contribute to the regulation of bone marrow and the production of blood-forming cells. Chemokines play a role in directing immune cells to sites of infection or inflammation. Additionally, there are many other cytokines that perform various functions in immune regulation that do not fall under these categories.

Cytokines can affect the cells that secrete them, neighboring cells, or in some cases, distant cells. It is known that a single cytokine affects several different types of cells, and different types of cells secrete the same cytokine.

Antibodies

They are protective proteins produced by the immune system through plasma cells (a type of white blood cell) in response to antigens (a substance that stimulates the body to produce a specific immune response).Each antibody binds to only one specific antigen, and the purpose of this binding is to help destroy the antigen. Some antibodies directly destroy antigens, while others assist in facilitating the destruction of the antigen by white blood cells.

Each antibody contains four peptides (peptides are chains composed of two or more amino acids) and the antibody consists of two heavy chains (each weighing about 50 kilodaltons) and two light chains (each weighing about 25 kilodaltons), these chains are linked by disulfide bonds, resulting in the formation of a distinctive Y-shaped structure with a total molecular weight of approximately 150 kilodaltons.

There are five main types of antibodies, each having a distinct structure and function: **Immunoglobulin G (IgG), Immunoglobulin M (IgM), Immunoglobulin E (IgE), Immunoglobulin A (IgA), and Immunoglobulin D (IgD).**

-Immunoglobulin G (IgG): This type of antibody was discovered in 1946 and is considered the most abundant among antibodies in human serum, constituting about 70-75% of the total antibodies present in the blood. It also makes up about 10-20% of plasma proteins. Its concentration in serum is approximately 12 g/L, and its molecular weight is about 146 kilodaltons.

This type of antibody is largely responsible for protecting the newborn during the first months of life because it is the only type of antibody that can cross the placenta in humans. Immunoglobulin G (IgG) is primarily produced during the secondary immune response to pathogens.

The human body contains four subclasses of immunoglobulin G (IgG1, IgG2, IgG3, IgG4), and these subclasses were discovered in the 1960s. IgG1 represents about 65% of the total immunoglobulin G, making it the largest type of immunoglobulin G. IgG2 constitutes between 20-25% of the total immunoglobulin G and is the second largest type of immunoglobulin G. The proportion of IgG3 is about 5-10% of the total immunoglobulin G, while IgG4 usually accounts for less than 4% of the total immunoglobulin G.

IgG1 is primarily responsible for the thymus-mediated immune response against proteins and polypeptide antigens. IgG2 deficiency is associated with recurrent infections in the airway and respiratory tract in infants, while IgG3 plays a major role in immune responses against protein or polypeptide antigens. Recent studies have also shown elevated serum levels of IgG4 in patients suffering from pancreatitis and cholangitis.

-Immunoglobulin M (IgM): It is the first type of antibody that the body produces in response to infection or immunization. It is the largest type of antibody and is also called (acute-phase antibodies) because it is produced rapidly in the early stages of the immune system's response. The main source of IgM production is the plasma cells found in the human spleen. IgM represents about 5% of the immunoglobulins in the serum .

Immunoglobulin M (IgM) appears early in the course of infection, and sometimes these antibodies appear again, but to a lesser degree. The appearance of these antibodies in the serum of an adult patient is considered an indication of a recent infection, but when they appear in the serum of newborns, they indicate an infection within the uterus .

-Immunoglobulin E (IgE): It is a type of antibody, discovered in 1966, and is considered the last of the five classes of human antibodies to be discovered. It has a molecular weight of 188 kilodaltons and a serum concentration of approximately 0.0003 g/L. Immunoglobulin E is the least common immunoglobulin in the blood serum and is present in healthy individuals at concentrations about 10,000 times lower than those of Immunoglobulin G (IgG). IgE levels increase significantly in type I hypersensitivity reactions, including allergic diseases such as parasitic infections like schistosomiasis.

When the human body suffers from allergies, the immune system overreacts to allergens by producing immunoglobulin E (IgE), which then binds to cells that release chemicals, causing an allergic reaction. This reaction sometimes causes symptoms in the nose, mouth, lungs, or skin.

-Immunoglobulin A (IgA): One of the five main immunoglobulins, it plays a central role in maintaining mucosal balance in the digestive, respiratory, and genitourinary systems and acts as the dominant antibody in immunity in this role. It is considered the second most abundant type of immunoglobulin in the body and thus has a crucial role in protection against antigens.

Immunoglobulin A (IgA) is distributed in the blood and is found in tears, saliva, breast milk, and fluids secreted from the lining of the airways, lungs, and digestive system.

There are two types of Immunoglobulin A (IgA), which are (IgA1) and (IgA2). Both are dense glycoproteins and each plays a different role in the body. (IgA1) is found on mucosal surfaces and constitutes about 80% of the total Immunoglobulin A (IgA), while (IgA2) is mainly found in the colon.

-Immunoglobulin D (IgD): It is a small molecule with a molecular weight of 184 kilodaltons , discovered in 1964. It is found in a small amount in serum, ranging from 0.03 mg/ml. Immunoglobulin D (IgD) constitutes about 0.25% of the total immunoglobulins in serum. The half-life of immunoglobulin D (IgD) is similar to that of immunoglobulin E (IgE), with a half-life of 2.8 days. Immunoglobulin D consists of two heavy delta (δ) chains and two light chains. The heavy and light chains are divided into variable and constant regions, in addition to disulfide bonds that link the chains together. Disulfide bonds are present within the chain, dividing each chain into regions known as domains. The light chains contain two domains, one variable and one constant, while the heavy chains contain four domains: one variable and three constant domains.

It has been found that immunoglobulin D (IgD) binds to basophils and mast cells and stimulates these cells to produce antimicrobial factors, thereby participating in the respiratory immune defense in humans.
Immunoglobulin D (IgD) primarily functions as an antigen receptor on B-lymphocytes (a type of white blood cell) and may be involved in regulating the function of B-lymphocytes when they encounter an antigen.

Dopamine Receptors

Dopamine is a neurotransmitter and a monoamine hormone of the catecholamine type. There are two amino acids involved in the production of dopamine: tyrosine and phenylalanine. Enzymes in the body convert tyrosine into dopamine through a mechanism that transforms the amino acid tyrosine into an intermediate substance called (L-DOPA), which then converts into dopamine. All of this occurs within the nerve cells in the midbrain region. Studies have shown that increased intake of these proteins leads to an increased rate of dopamine in the body.

It plays a role in the brain by helping neurons exchange messages with each other as well as between the brain and the rest of the body. Since dopamine is a neurotransmitter, it plays an important role in a range of functions including attention, behavior, movement, learning, memory, activity, motivation, wakefulness, mood, sleep, and happiness.

As a hormone, dopamine works to:

1-At low doses, dopamine causes blood vessels to dilate and relax, while at high doses it causes them to contract.

2-It reduces insulin production in the pancreas.

3-It decreases the activity of lymphocytes in the immune system.

4-It increases sodium levels in the body and promotes urine excretion.

Dopamine receptors are classified as G protein-coupled receptors (G proteins are guanine nucleotide-binding proteins) and play an important role in regulating motor activity and several neurological disorders such as schizophrenia, Alzheimer's disease, Parkinson's disease, bipolar disorder, and attention deficit hyperactivity disorder.

Dopamine receptors are found in the central nervous system, especially in the dentate gyrus of the hippocampus in the temporal lobe of the brain, as well as in the subventricular zone (located near the lateral ventricle in the brain). Dopamine receptors are also present in the periphery, with a higher presence observed in the kidneys and blood vessels.

There are five types of dopamine receptors, which are D1, D2, D3, D4, and D5. Each receptor has a different function and is located in different sites.

The function of each dopamine receptor is as follows:

D1: Memory, concentration, emotion regulation, kidney functions, movement.

D2: Movement, concentration, sleep, memory, learning.

D3: Perception, emotion regulation, concentration, sleep.

D4: Perception, concentration, emotion regulation, memory, feeling of fear, sleep.

D5: Concentration, decision making, perception, renin secretion.

2- Structural proteins

Structural proteins are a group of proteins responsible for functions ranging from cell shape and movement to supporting major structures such as bones, cartilage, muscles, and hair. Structural proteins include a group of proteins such as collagen, elastin, and keratin.

Elastin

It is one of the main structural proteins in the body, highly cross-linked and more abundant in tissues, imparting elasticity to tissues such as blood vessels, elastic ligaments, skin, and lungs. Elastin has hydrophobic properties and is insoluble; it can also resist acids and alkalis. It plays a vital role in elastic fibers and is a fibrous protein similar to collagen, with more than 75% of the elastin sequence consisting of only four non-polar amino acids: glycine, alanine, valine, and proline.

Elastin is composed of the highly cross-linked protein tropoelastin, which weighs about 65 kilodaltons. Elastin is a long-lived protein due to its extensively cross-linked structure, degrading slowly in healthy tissues, with a half-life of about 70 years.

Elastin is synthesized early in human life , and elastin synthesis in humans significantly slows down by the age of forty. Elastin is produced in cells called fibroblasts, and vitamin C plays an important role in the process of elastin synthesis. Zinc also helps in the process of elastin synthesis.

Elastin enhances the skin's ability to stretch and return to its original state, preventing sagging and reducing the appearance of wrinkles. It also helps restore skin elasticity after weight loss. Additionally, elastin plays an important role in skin cell regeneration, which helps improve the skin's appearance and make it look younger. Elastin also enhances the elasticity of blood vessels, contributing to supplying the skin with the necessary nutrients.

Elastin is added to a wide range of anti-aging products because the process of elastin synthesis is directly related to the appearance of skin wrinkles and other aging processes.

3- Contractile Proteins

Contractile proteins are responsible for the contraction of muscle fibers and are also called motor proteins, and consist of thin filaments (actin, troponin, tropomyosin) and thick filaments (myosin) (with titin providing structural support), that regulate the strength and speed of heart and muscle contractions. Their expression is regulated by physiological signals specific to different types of cells.

Contractile proteins are widely distributed in nature and are found in almost every living cell, not only in motor organs. They have recently been isolated from the brain, thyroid gland, pancreas, liver, and others.

The contractile proteins found in animal and plant cells share many fundamental properties, including enzymatic, physical, and chemical characteristics.

Actin

It is a major structural protein that forms the cytoskeleton of cells and is involved in functions such as cell division, migration, and vesicle transport. Actin exists either in a monomeric form known as G-actin (Globular Actin) or in a filamentous form (F-actin) (Fibrous Actin), and each actin subunit binds either to an adenosine triphosphate (ATP) molecule or an adenosine diphosphate (ADP) molecule.

Actin constitutes 10% of the total cellular protein in muscle cells, while in non-muscle cells it is less, ranging between 1% and 5% of the total cellular protein.

Actin filaments organize themselves into dynamic bundles or networks that play essential roles in various cellular processes. Actin bundles are cross-linked and tightly packed in parallel rows, whereas in actin networks, actin filaments are loosely cross-linked and form three-dimensional meshworks.

The main functions of actin filaments include the following:

-Forming a dynamic cytoskeleton to provide structural support to cells.

-Supporting muscle contractions through the sliding of actin filaments alongside myosin filaments.

-Supporting and enabling cell movement.

Titin

It is a large, flexible filamentous protein with a molecular weight of 3 million Daltons and consists of 27,000 amino acids, also called (connectin). It is the largest protein in mammals, located in the sarcomere (the main contractile unit of muscle fibers in skeletal muscle) in striated muscles. It forms a flexible filament that extends along the muscle fibers in both cardiac and skeletal muscles and is important for producing passive force during muscle contraction. The adult human body contains about 0.5 kilograms of this protein, which is more than 1.5 micrometers in length and 3-4 nanometers in width. This protein makes up about 10% of muscle mass, making it the third most abundant protein in muscles after actin and myosin.

Another important role of titin is that it acts as a regulatory node that integrates and possibly coordinates various signaling pathways, especially in cardiac muscle cells. In 1979, Kuan Wang, Janela McClure, and Ann Tu provided a comprehensive description of titin.

4- Transport Proteins

These are a type of proteins that work to transport substances across biological membranes. These proteins are located within the membrane itself, where they form a channel or transport mechanism that allows substances to move from one side to the other.

Transport proteins are considered important for the growth and life of all living organisms. There are various types of transport proteins. Transport proteins are also essential for communication between cells in the human body, especially in interactions between nerves and the transport of monoamines.

These proteins contribute to the reabsorption of neurotransmitters in the brain, ensuring the continuation of neural transmission by reducing their availability in the synaptic cleft.

Impaired function of transport proteins can lead to decreased levels of monoamine neurotransmitters, which are associated with conditions such as depression.

The substances transported by these proteins include ions such as potassium, sodium, and calcium, in addition to sugars like glucose, amino acids, carrier molecules, and other diverse materials.

In general, transport proteins perform two types of transport: passive transport and active transport.

Passive transport is the most common, occurring mostly through diffusion along the concentration gradient from high to low concentration and does not require energy. Examples include the diffusion of neurotransmitters such as acetylcholine across the synapse or the neuromuscular junction. In addition to diffusion, passive transport includes facilitated diffusion and osmosis.

Facilitated diffusion occurs when molecules such as amino acids or glucose move from a high concentration to a low concentration with the help of carrier proteins or pores in the membrane. Osmosis is the movement of water molecules from an area of high water concentration to an area of low water concentration through a selectively permeable membrane.

Active transport is the movement of molecules or ions across the cell membrane from an area of low concentration to an area of higher concentration, against the concentration gradient. It is an energy-dependent process. There are two types of active transport: primary active transport, which relies on the use of adenosine triphosphate (ATP), and secondary active transport, which depends on the use of an electrochemical gradient.

Albumin

It is a globular protein with a molecular weight ranging between 66 - 69 kilodaltons, consisting of 585 amino acids. It is the most abundant protein in plasma, representing more than half of the total protein content in plasma (accounting for 60% of the protein volume present in plasma). Its level ranges between 3.5g/dL to 5g/dL in a healthy human. It has a long half-life, reaching up to 19 days.

Liver cells produce albumin and secrete it into the bloodstream at a rate ranging from 10 to 15 grams daily. Very little albumin is stored in the liver, and most of it is rapidly secreted into the bloodstream. However, the presence of high concentrations of albumin in other fluids such as urine often indicates an underlying disease.

The hepatic synthesis of albumin is affected by several factors that lead to the inhibition of albumin production. The most important of these factors are malnutrition (fasting leads to a decrease in albumin by one-third within 24 to 48 hours of starting fasting), exposure to hepatotoxins, infections, and exposure to high osmotic pressure.

In humans, albumin regulates the colloid osmotic pressure in plasma and transports various substances called ligands.

These ligands transported by blood albumin include ions, bilirubin, fatty acids, hormones, as well as drugs such as methadone, alfentanil, warfarin, furosemide, propranolol, methotrexate, thiopental, and others.

Albumin also performs other functions such as maintaining blood pH and plays a significant role as an antioxidant and anti-inflammatory agent.

Albumin is also present in breast milk and has properties similar to those found in blood.

It is believed that it may not be produced by the mammary gland but rather transferred from the mother's bloodstream. Additionally, blood albumin is considered a source of amino acids for the breastfed infant.

5- Regulatory Proteins

These are a type of proteins that control gene expression by binding to specific sequences of DNA, thereby enhancing or inhibiting the transcription process. They are also called (transcription factors). These proteins play a key role in maintaining cell functions by regulating processes such as cell growth, differentiation, and response to environmental stimuli. They also manage the rate of transcription of genetic information from deoxyribonucleic acid (DNA) to messenger ribonucleic acid (mRNA), which affects various biological processes.

Regulatory proteins are essential for many functions in living organisms, including:

-Gene expression: Regulatory proteins have the ability to turn genes on or off, affecting the proteins produced by the cell.

-Enzyme regulation: Regulatory proteins can increase or decrease enzyme activity, as these proteins accelerate biochemical reactions.

-Cell cycle control: Regulatory proteins help regulate the cell cycle and ensure proper cell division and proliferation.

-Signal transduction: Regulatory proteins play a major role in transmitting signals from outside the cell to inside it.

-Regulatory proteins: also play a crucial and important role in regulating essential nutrients in the human body, helping to organize vital processes such as digestion, metabolism, immune functions, energy production, and others.

Regulatory proteins also play a vital role in many metabolic functions, including energy regulation (these proteins help maintain energy balance in the body by adjusting metabolic rates), appetite control (regulatory proteins affect hunger and satiety, thereby influencing food intake), and some regulatory proteins are essential in synthesizing the essential nutrients required for various bodily functions.

In addition to their primary functions, regulatory proteins can be potential targets for treating different diseases, such as cancer, where regulatory proteins can be engineered or modified to prevent the uncontrolled cell division characteristic of the disease.

Examples of regulatory proteins include: insulin hormone, glucagon hormone, ferritin, leptin.

Insulin

Insulin is a protein (peptide hormone) composed of two chains: chain A and chain B. Chain A contains 21 amino acids, while chain B contains 30 amino acids. They are linked together by a disulfide bond (disulfide bridge), and its molecular weight is approximately 5802 Daltons. Insulin is derived from a relatively inactive prohormone molecule known as proinsulin, which consists of 74 amino acids. Insulin was first discovered by the Canadian surgeon Frederick Banting and his student Charles Best in 1921.

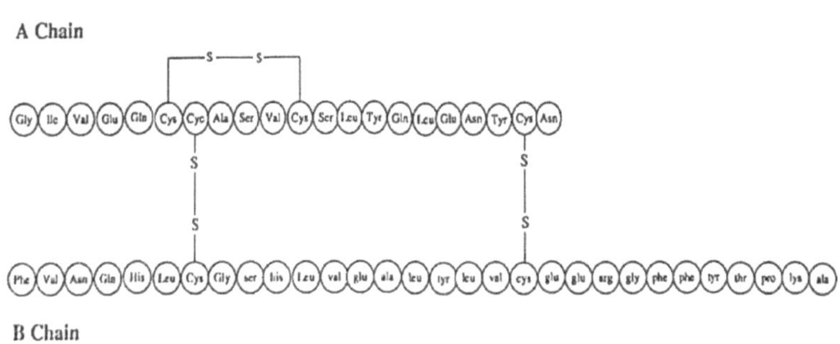

Diagram showing the structure of insulin

Insulin is synthesized in beta cells (specialized endocrine cells located within the islets of Langerhans in the pancreas), where ribosomes translate messenger RNA (mRNA) in the endoplasmic reticulum to form preproinsulin (with a molecular weight of 11,500 Daltons).

Then, preproinsulin is cleaved in the endoplasmic reticulum to form proinsulin (with a molecular weight of 9,000 Daltons).

After that, proinsulin is cleaved in the Golgi apparatus to form insulin, which is stored in large dense-core secretory granules known as beta granules.

These large dense-core granules mainly consist of a large insulin crystal chelated with zinc (Zn^{2+}). It is estimated that each beta granule contains more than 250,000 insulin molecules.

The half-life of beta granules is about 3 to 5 days, and if they do not undergo cellular exocytosis (insulin release), they are degraded by autophagy.

Insulin is considered the key regulator in maintaining glucose balance, metabolism, and cell growth. Elevated blood sugar stimulates beta cells to secrete a small amount of insulin (less than 5%) within one to ten minutes after eating.

The systemic circulation distributes insulin to the main target cells, which are liver cells, thereby promoting glucose storage through glycogen synthesis, as well as skeletal muscle cells and fat cells to stimulate glucose uptake, resulting in the blood glucose level dropping to its normal range.

Mechanism of Insulin Release from Beta Cells in the Pancreas

A beta cell primarily contains ion channels, including the ATP-sensitive potassium ion channel and the voltage-gated calcium channel. The beta cell also contains other organelles such as the nucleus, endoplasmic reticulum, Golgi apparatus, mitochondria, and a glucose transporter protein called Glucose Transporter 2 (GLUT2). Additionally, inside the beta cell, there are secretory vesicles that contain insulin.

When the glucose concentration in the blood is normal, in other words, not elevated, a state called the resting state, the ATP-sensitive potassium ion channel is open, allowing potassium ions to flow out of the beta cell. This makes the inside of the cell more negative compared to the outside (since potassium ions carry a positive charge), creating a voltage difference across the beta cell membrane. The inside of the beta cell is at a potential of -70 millivolts compared to the outside, making the cell membrane hyperpolarized.

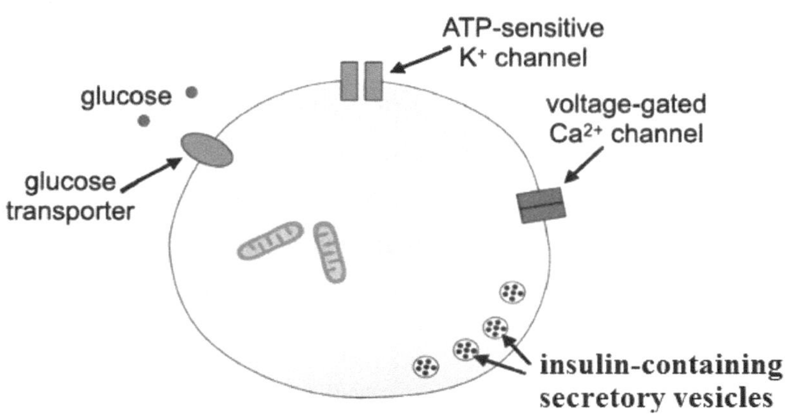

Beta cell in resting state

When the concentration of glucose in the blood increases, this condition is called glucose stimulation. More glucose enters the beta cell through facilitated diffusion via a glucose transporter called (Glucose Transporter 2) (GLUT2), where this glucose is used by the mitochondria for respiration and the production of adenosine triphosphate (ATP).

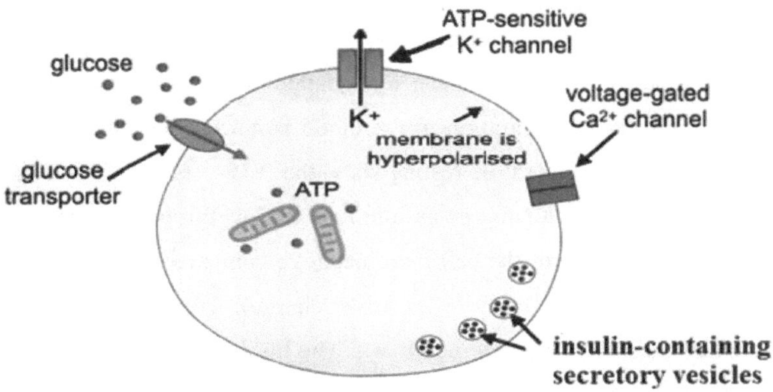

Beta cells in a state of glucose stimulation

The increase in the amount of ATP inside the beta cell leads to the closing of potassium ion channels, preventing potassium ions from diffusing out of the cell. This causes the membrane potential to become more positive (due to the accumulation of potassium ions inside the cell), leading to a decrease in the cell surface membrane potential to (-30 millivolts).

As a result, the membrane depolarization occurs, which stimulates voltage-gated calcium channels to open, allowing calcium ions to enter the cell by diffusion. This increases the concentration of calcium ions inside the cell, which in turn stimulates secretory vesicles containing insulin to move to the cell surface and fuse with the cell membrane. Then they open, releasing insulin into the bloodstream, which is transported throughout the body.

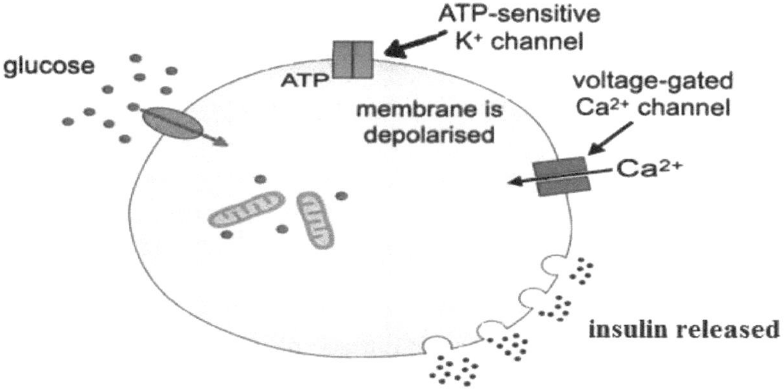

Beta cell in the state of releasing insulin into the bloodstream

Glucagon hormone

It is a peptide hormone composed of 29 amino acids, with a molecular weight of 3485 Daltons. The half-life of glucagon in plasma ranges between 5 and 10 minutes and it is broken down in the liver. This hormone is mainly secreted by the alpha cells in the islets of Langerhans in the pancreas, where it is derived from proglucagon (a protein composed of 160 amino acids that can be converted into several related peptide hormones). Proglucagon is produced in pancreatic alpha cells and, to a lesser extent, in neurons in the brainstem and hypothalamus.

The history of glucagon began in 1923 when Charles Kimball and John Murlin identified a circulating factor that affected glucose balance in the body, with an effect opposite to that of insulin.

Glucagon is secreted in response to low blood sugar, exercise, and stress, leading to an increase in blood glucose concentration by stimulating glucose production and glycogen breakdown in the liver.

Glucagon hormone secreted by the alpha cells of the pancreatic islets of Langerhans plays a very important role in maintaining glucose balance, as its mechanism of action is opposite to that of insulin. Glucagon acts as a glucose-mobilizing hormone, and in accordance with these opposing mechanisms, high plasma glucose concentrations that stimulate insulin secretion from pancreatic beta cells, inhibit glucagon secretion, while low plasma glucose concentrations represent one of the strongest stimuli for glucagon secretion.

Excessive production of glucagon from the alpha cells of the pancreatic islets of Langerhans is also a characteristic feature of glucagonoma (a rare type of tumor affecting the alpha cells in the pancreas). Therefore, serum glucagon is considered a useful tumor marker for diagnosing this tumor.

Glucagon plays a physiological role not limited to regulating blood glucose levels but also includes amino acid metabolism. Thus, there appears to be an axis between the liver and alpha cells, where glucagon regulates amino acid metabolism in the liver, and amino acids regulate the growth and secretion of alpha cells.

Ferritin

It is a large, hollow, and complex globular protein, with a mass of about 474 kilodaltons. Ferritin consists of a spherical shell made up of 24 subunits (each ferritin subunit folds to form a four-helix bundle with a short fifth helix near the carboxyl terminal) with a central core that can contain up to 4,500 iron atoms (iron is stored inside ferritin in the form of a mineral called ferrihydrite). The inner diameter of ferritin is 8 nanometers and the outer diameter is 12 nanometers. There are two functionally and genetically distinct ferritin subunits: the light chain ferritin and the heavy chain ferritin (ferritin-L and ferritin-H).

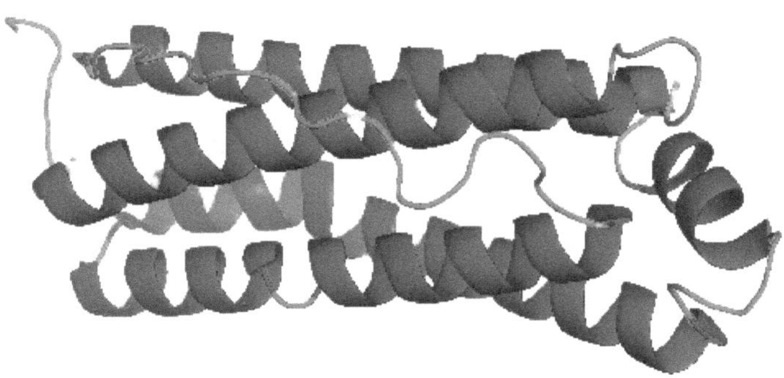

Crystallographic structure of mitochondrial ferritin

Ferritin is the oldest known protein involved in iron metabolism. It was first described by the German pharmacist Oswald Schmiedeberg in 1894, who observed the presence of an iron-rich component in horse liver. It was extracted from horse spleen in 1937 by the Czech biologist Vilém Laufberger, who proposed that there must be a substance that acts as an iron reservoir in living organisms.

Ferritin is a multifunctional protein with potential roles in cell proliferation, angiogenesis, immune suppression, and iron transport. It also has antioxidant and anti-inflammatory functions. Additionally, ferritin levels are considered a diagnostic indicator of iron deficiency anemia. Since ferritin levels are elevated in the serum of many cancer patients and these elevated levels correlate with disease severity, it has been suggested to use serum ferritin as a tumor marker in the diagnosis of malignant tumors.

Both hemoglobin and ferritin are iron-related proteins in the blood. Hemoglobin carries oxygen, which binds to the heme component coordinated with iron ions, to body tissues. Ferritin, on the other hand, is the main iron-binding protein that stores and releases iron in tissues, regulating iron levels in organs such as the liver, spleen, and bone marrow.

It has been observed that as cancer progresses, ferritin concentration often increases while hemoglobin levels decrease. This phenomenon differs from iron deficiency anemia, where both ferritin and hemoglobin decrease simultaneously .

Leptin

It is a peptide hormone secreted by white adipose tissue and the small intestine, especially in intestinal cells, and is also found in the stomach, mammary glands, ovaries, placenta, skeletal muscles, and others. This

protein consists of 167 amino acids and is called the (satiety hormone or severe hunger hormone).
In 1994, a team of researchers led by geneticist Jeffrey Friedman discovered the hormone leptin, and it was named after the Greek word (leptos), which means thinness.
The structure of leptin resembles the structure of pro-inflammatory cytokines found throughout the body.

Leptin performs its functions by binding to leptin receptors located on the cell surface, where these receptors are found in various tissues including neural, cardiac, hepatic, pancreatic, and intestinal regions.
The amount of leptin in the blood increases proportionally with the increase in the amount of adipose tissue in the body, as there is a close relationship between leptin levels in the bloodstream and the amount of adipose tissue.

Leptin deficiency or resistance is associated with dysregulation in cytokine production, malnutrition, autoimmune disorders, increased susceptibility to infections, and inflammatory responses.

Leptin deficiency leads to a range of clinical symptoms including morbid obesity, increased hyperphagia, reduced satiety, hyperinsulinemia, persistent food-seeking behavior, recurrent bacterial infections, fatty liver disease (hepatic steatosis), dyslipidemia (a metabolic disorder characterized by abnormally high or low levels of any or all lipids such as cholesterol, triglycerides, and phospholipids), and hypogonadotropic hypogonadism (a form of gonadal failure caused by a problem in the pituitary gland or hypothalamus, resulting in failure to produce gonadal hormones such as estrogen and testosterone).
Leptin resistance, especially its effects on appetite suppression and body

weight reduction, is linked to hyperleptinemia. Hyperleptinemia and leptin resistance are components of common obesity, as the direct correlation between serum leptin concentrations and body fat percentage serves as evidence of this association.

Obese individuals have higher serum leptin levels as well as higher leptin mRNA content in adipose cells compared to individuals with normal weight.

Additionally, serum leptin levels and leptin mRNA content in adipose cells decrease with weight loss.

Leptin resistance mechanisms may be related to a defect in the transport of leptin across the blood-brain barrier or to problems in the intracellular signaling mechanisms located downstream of the leptin receptor.

There are many diseases associated with hyperleptinemia, including Rabson-Mendenhall syndrome, non-alcoholic fatty liver disease, depression, neurodegenerative disorders, and food addiction.

Chapter

Five

Protein Synthesis in the Cell

The Role of Deoxyribonucleic Acid (DNA) and Ribonucleic Acid (RNA) in Protein Production

Deoxyribonucleic acid (DNA) is considered the fundamental molecule of life, as it contains the essential information that the cell needs to synthesize proteins and replicate. Therefore, DNA is regarded as a repository of the necessary information for any cell to perform its functions.

The current structure of DNA was discovered in 1953 by James Watson and Francis Crick. DNA has a double helix structure, which is of great importance. DNA consists of four nitrogenous bases: adenine (A), guanine (G), thymine (T), and cytosine (C). The DNA sequence appears as follows: " ATTGCGTA"

In the double helix structure of DNA, each base pairs with its complementary base, where adenine (A) pairs with thymine (T), and similarly, guanine (G) pairs with cytosine (C). Additionally, DNA molecules are characterized by their remarkable length. DNA is divided into small parts and tightly coiled into structures called chromosomes. Humans have 23 pairs of chromosomes, and these chromosomes are further divided into smaller coding segments called genes. The 23 pairs of chromosomes contain approximately 20,000 to 25,000 genes, each with its specific function.

The unit of measurement for deoxyribonucleic acid is the kilobase (abbreviated as kb), where one kilobase equals 1000 bases.

The cases in which there is a change in the genetic code are called mutations. There is an intermediate coded language in the ribonucleic acid (RNA) sequence that translates the gene message into a sequence of amino acids for the protein, where the protein determines the trait. This is called the (central dogma of molecular biology).

Ribonucleic acid (RNA) resembles deoxyribonucleic acid (DNA) as both contain nitrogenous bases attached to a sugar-phosphate backbone, but there are functional and structural differences that distinguish RNA from DNA.

Structurally, deoxyribonucleic acid (DNA) consists of two strands and contains thymine (T) that pairs with adenine (A) in the double strand. The nucleotides of DNA also include a sugar (deoxyribose).

On the other hand, ribonucleic acid (RNA) consists of a single strand and contains uracil (U) that pairs with adenine (A). The nucleotides of RNA include a sugar (ribose), as shown in the following figure.

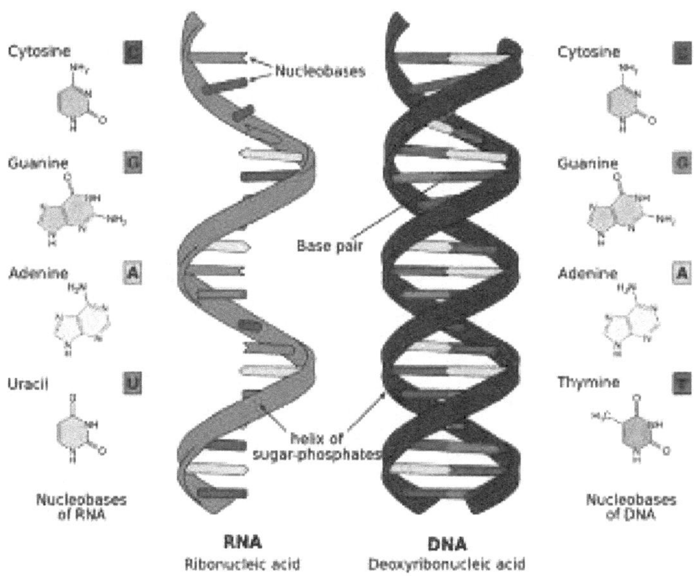

DNA vs RNA - Similarities and Differences

Functionally, deoxyribonucleic acid (DNA) serves as a storage for protein-coding information, while ribonucleic acid (RNA) uses this information to enable the cell to produce that protein.

Protein synthesis

Protein synthesis is the process of creating new functional peptides or renewing existing functional peptides. The process of protein synthesis is considered an essential part of cellular regulation and has profound effects at the tissue, organ, and whole-body levels.

Protein synthesis can be stimulated or inhibited in response to internal and external stimuli, allowing for appropriate adaptation and responses to various stresses.

Newly synthesized proteins function as enzymes, hormones, antibodies, and transporters, and they can also assist in vital functions such as acid-base balance, fluid and electrolyte balance, immune defense, energy production, cell growth, maintenance, and repair.

Therefore, the process of protein synthesis is subject to precise regulation, and disruptions in protein synthesis can lead to an increased likelihood of disease development and progression, and even death.

The process of protein synthesis in the cell consists of two main steps:

1-Transcription

2-Translation

Transcription process

The first step in the protein synthesis process is transcription, which takes place in the nucleus by transferring the information contained in deoxyribonucleic acid (DNA) to a messenger ribonucleic acid (mRNA) molecule. Since DNA and RNA are chemically and structurally similar, DNA can serve as a direct template for synthesizing RNA through complementary base pairing.

During transcription, the DNA of the gene acts as a template for complementary base pairing, and an enzyme called RNA polymerase II (RNAP II) catalyzes the formation of a precursor messenger RNA (pre-mRNA) molecule, which is then processed to form mature messenger RNA, as shown in the figure.

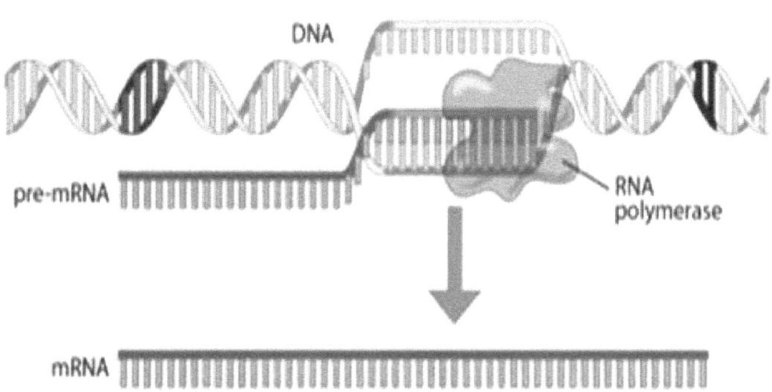

Transcription process

The resulting messenger ribonucleic acid (mRNA) is a single-stranded copy of the gene that will later be translated into a protein. After transcription is complete, the mature mRNA molecule leaves the nucleus through the nuclear pores and enters the cytoplasm, where translation takes place.

Translation process

The messenger ribonucleic acid (mRNA) transcribed leaves the nucleus and moves to the cytoplasm, where it binds to the ribosome (the ribosome is the organelle responsible for protein synthesis). Transfer ribonucleic acid (tRNA) molecules complementary to the messenger ribonucleic acid (mRNA) arrive at the ribosome. When the messenger ribonucleic acid (mRNA) reaches the ribosome, it recruits the complementary transfer ribonucleic acid (tRNA) to the base sequence of the messenger ribonucleic acid.

Transfer ribonucleic acid (tRNA) molecules contain a specific region called the anticodon (the codon is a triplet of three nitrogenous bases that can align according to the genetic code to form a DNA sequence), which is complementary to a codon on the messenger ribonucleic acid (mRNA) molecule. The transfer ribonucleic acid (tRNA) molecule also carries amino acids. There are at least 20 types of transfer ribonucleic acid (tRNA) in cells, each with a unique anticodon and a corresponding amino acid. The bases on the messenger ribonucleic acid (mRNA) are read in triplets and encode a specific amino acid.

The translation process can be explained in detail as follows:

The first transfer ribonucleic acid (tRNA) (which always carries the amino acid methionine) forms hydrogen bonds with the first codon (start codon) on the messenger ribonucleic acid (mRNA).

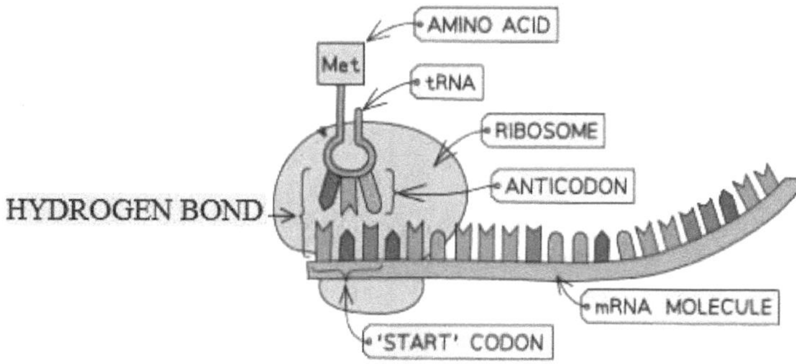

Then the second transfer ribonucleic acid (tRNA) (which carries the second amino acid) binds to the second codon on the messenger ribonucleic acid (mRNA). After that, a peptide bond forms between the amino acids present on the transfer ribonucleic acid (tRNA).

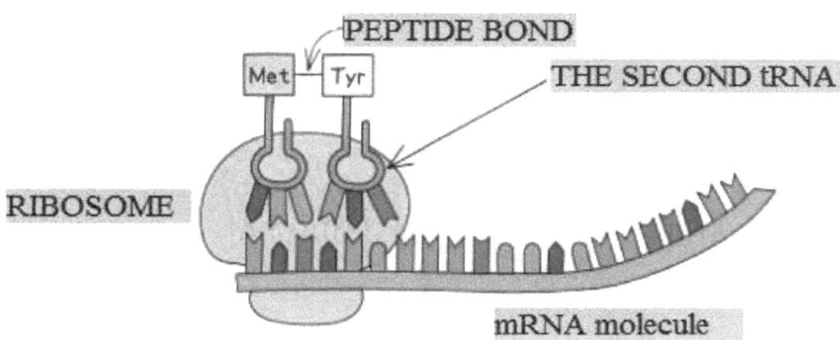

The ribosome moves along the messenger RNA (mRNA) in the 5 to 3 direction and reads the next codon. Then, the third transfer RNA (tRNA) (which carries the third amino acid) binds to the complementary codon, and the first transfer RNA (tRNA), which does not contain an amino acid, is released.

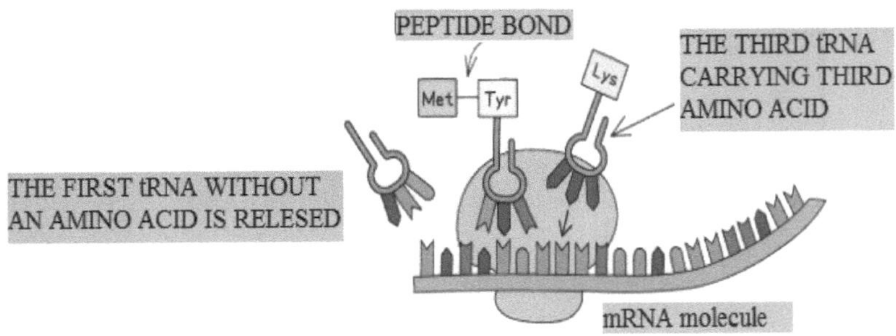

The ribosome continues to read the messenger RNA (mRNA) molecule, forming a polypeptide chain until it reaches the stop codon.

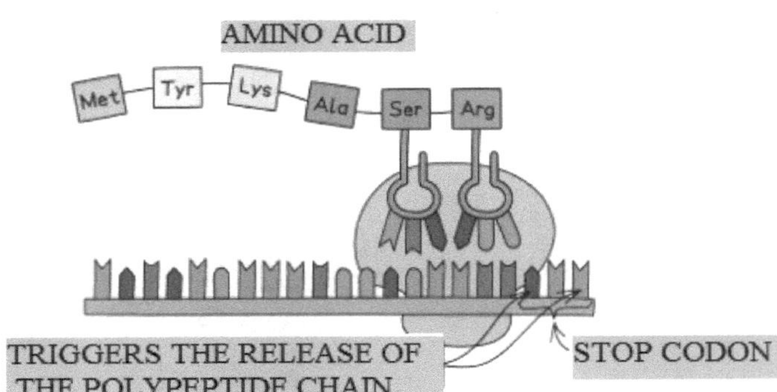

Post-translational modifications

These are common typical biochemical reactions that covalently link peptide chains or chemical groups or carbohydrates or lipids to the amino acids of the target molecule during or after its translation.

The process of post-translational modification of proteins is an important and fundamental means of regulating their functions. About 50% to 90% of proteins in human cells undergo different types of post-translational modifications.

In most known proteins, post-translational modifications occur, where almost all amino acids can be altered by one or more of these reactions. Modified proteins acquire rare amino acids that may greatly affect the protein's structure and function.

Thus, proteins are diversified by changing their structure, location, interactions, regulation, and function, which leads to impacts on all biological aspects of the cell.

Post-translational modifications are particularly important in the study of complex, multigenic diseases such as cancer, cardiovascular diseases, and diabetes.

More than 650 types of protein modifications have been identified so far, including some such as protein phosphorylation, glycosylation, methylation, and acetylation.

Protein Phosphorylation

Protein phosphorylation is considered the most common post-translational modification and is a reversible modification. It occurs through the covalent attachment of a phosphate group to an amino acid, where a phosphate group is transferred from adenosine triphosphate (ATP) to the hydroxyl group in the side chain of amino acid residues via a catalytic reaction by kinase enzymes (all kinases require the presence of a divalent metal ion such as Mg^{2+}or Mn^{2+} for their catalytic activity). The phosphate group covalently binds to residues of serine (Ser), threonine (Thr), or tyrosine (Tyr).

The phosphorylated protein can be returned to its non-phosphorylated state by the enzyme protein phosphatase, which catalyzes the cleavage of the phosphoester bond through hydrolysis, as shown in the following figure.

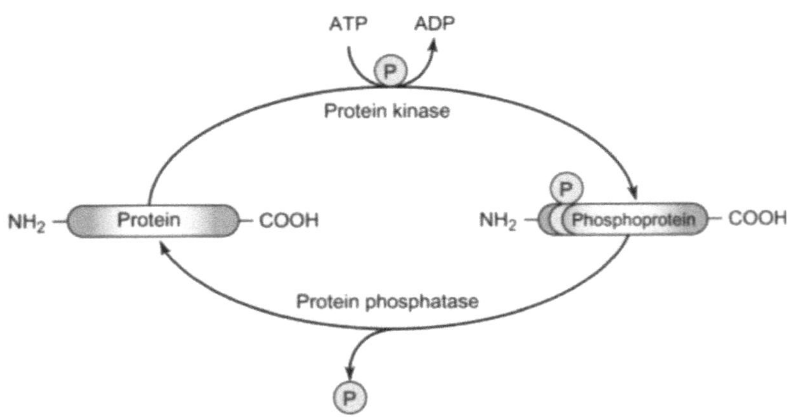

Protein Phosphorylation

This reversible transition based on the activity between the non-phosphorylated state and the phosphorylated state represents one of the most common molecular mechanisms for transmitting physiological signals to control cellular functions.

The addition of phosphate groups to proteins can lead to conformational changes that alter biochemical and cellular functions such as modifying enzyme activity, cellular localization, or molecular binding.

In 1906, protein phosphorylation was first reported by Phoebus Levene when he discovered phosphate in vitellin protein.

Protein phosphorylation plays a critical role in key cellular functions such as transcription, replication, cell movement, metabolism, programmed cell death, response to environmental stress, and immune system activity.

Research has proven that any dysfunction in phosphorylation pathways may contribute to the development of many diseases, including cancer, cardiovascular diseases, Parkinson's disease, and Alzheimer's disease.

Glycation

It is a reaction in which sugars bind to proteins, leading to the formation of aggregates known as glycoproteins. Glycation is considered one of the most complex post-translational modifications in the cell. It is a reversible, enzyme-directed reaction. The glycation process occurs in several subcellular locations such as the endoplasmic reticulum, the Golgi apparatus, the cytoplasm, and the sarcolemma membrane (a thin, fragile, and stretchable membrane surrounding skeletal muscle fibers or cardiac muscle cells).

Glycation is an important process in many biological functions such as protein folding regulation, transport, signal transduction, protein-ligand interactions (biochemical reactions occurring when a small molecule (ligand) binds to a protein), protein solubility effects, cell-cell interactions, receptor activation, and cell interactions with the surrounding matrix environment.

Protein glycation occurs in two different forms, leading to the formation of O-linked and N-linked sugars. O-linked glycation is the attachment of monosaccharides to the hydroxyl groups of serine or threonine residues in nascent proteins, as shown in the figure.

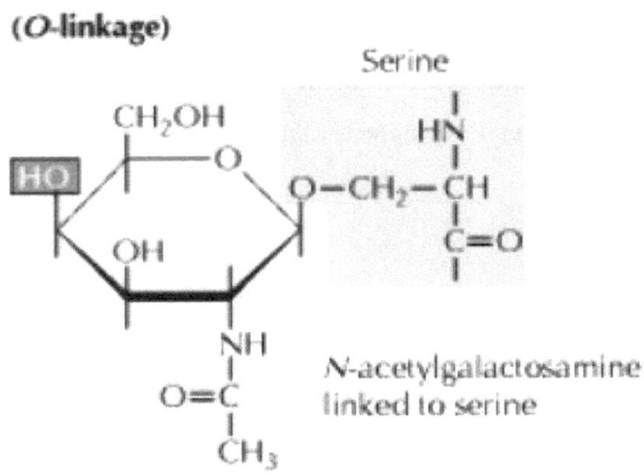

Diagram of O-Linked Glycosylation

While nitrogen-linked glycosylation involves the attachment of monosaccharides to amide groups in asparagine residues, as shown in the following figure .

(*N*-linkage)

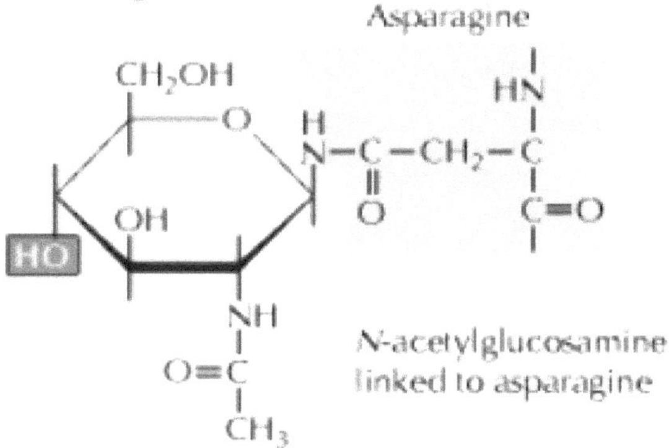

Diagram of N-Linked Glycosylation

Approximately 70% of secreted proteins undergo glycosylation, and these modifications are essential as they play a key role in directing these proteins to their final destinations, whether transported outside the cell or integrated into cellular organelles and the plasma membrane.

Deficiencies in glycosylation pathways lead to more than one hundred human diseases, as dysregulation of glycosylation has been associated with many human diseases such as diabetes, cancers, neurodegenerative diseases, liver fibrosis, atherosclerosis, and HIV infection.

Protein Methylation

Protein methylation is the process of adding a methyl group to the amino acid lysine or arginine in the protein, catalyzed by methyltransferase enzymes that transfer the methyl group from the methyl donor (S-adenosyl methionine) to the target amino acid.

This process plays an important role in epigenetic modifications that affect cancer biology.

Protein methylation is considered a significant post-translational modification, playing a key role in regulating various biological processes ranging from gene expression and signal transduction pathways to DNA repair and protein cycling.

Protein methylation occurs at the amino and carboxyl termini, as well as on nitrogen atoms in the side chains of arginine and lysine residues.

Protein methylation is as common as phosphorylation and glycosylation and can occur on different amino acids, including lysine, arginine, histidine, glutamate, and others. Arginine methylation in proteins is a common post-translational modification in eukaryotes, occurring in three distinct forms: monomethylarginine (MMA), symmetric dimethylarginine (SDMA), and asymmetric dimethylarginine (ADMA).

Disruption of protein methylation has been linked to a wide range of human disorders, such as cancer, cardiovascular diseases, inflammation, and neurodegenerative conditions. Consequently, there is increasing interest in targeting the enzymes that regulate methylation.

References

1- Wewer Albrechtsen, N. J., Kuhre, R. E., Pedersen, J., Knop, F. K., & Holst, J. J. (2016). The Biology of Glucagon and the Consequences of Hyperglucagonemia . Biomarkers in Medicine, 10(11), 1141–1151.https://doi.org/10.2217/bmm-2016-0090

2- Burnett, G., & KENNEDY, E. P. (1954). The enzymatic phosphorylation of proteins. The Journal of biological chemistry, 211(2), 969–980.

3- Cieśla, J., Frączyk, T., & Rode, W. (2011). Phosphorylation of basic amino acid residues in proteins: important but easily missed. Acta biochimica Polonica, 58(2), 137–148.

4- Lewandowski, J., Komur, A. A. & Sobańska, D. (2021). Structural and functional relationship of mammalian and nematode ferritins. BioTechnologia, 102(4), 457–471. https://doi.org/10.5114/bta.2021.111110

5- Shesh, B. P., & Connor, J. R. (2023). A novel view of ferritin in cancer. Biochimica et biophysica acta. Reviews on cancer, 1878(4), 188917. https://doi.org/10.1016/j.bbcan.2023.188917

6- Wang, W., Knovich, M. A., Coffman, L. G., Torti, F. M., & Torti, S. V. (2010). Serum ferritin: Past, present and future. Biochimica et biophysica acta,1800(8),760–769. https://doi.org/10.1016/j.bbagen.2010.03.011

7- Dutta, H., & Jain, N. (2023). Post-translational modifications and their implications in cancer. Frontiers in oncology, 13, 1240115. https://doi.org/10.3389/fonc.2023.1240115

8- CRICK F. H. (1958). On protein synthesis. Symposia of the Society for Experimental Biology, 12, 138–163.

9- Singh, R., Barden, A., Mori, T., & Beilin, L. (2001). Advanced glycation end-products: a review. Diabetologia, 44(2), 129–146. https://doi.org/10.1007/s001250051591

10- Knovich, M. A., Storey, J. A., Coffman, L. G., Torti, S. V., & Torti, F. M. (2009). Ferritin for the clinician. Blood reviews,23(3),95–104. https://doi.org/10.1016/j.blre.2008.08.001

11- Langlois d'Estaintot, B., Santambrogio, P., Granier, T., Gallois, B., Chevalier, J. M., Précigoux, G., Levi, S., & Arosio, P. (2004). Crystal structure and biochemical properties of the human mitochondrial ferritin and its mutant Ser144Ala. Journal of molecular biology, 340(2), 277–293. https://doi.org/10.1016/j.jmb.2004.04.036

12- Nestler, E. J., & Greengard, P. (1999). Protein phosphorylation is of fundamental importance in biological regulation. Lippincott Raven.

13- Boland, B. B., Brown, C., Jr, Alarcon, C., Demozay, D., Grimsby, J. S., & Rhodes, C. J. (2018). β-Cell Control of Insulin Production During Starvation-Refeeding in Male Rats. Endocrinology,159(2),895–906. https://doi.org/10.1210/en.2017-03120

14- Uchizono, Y., Alarcón, C., Wicksteed, B. L., Marsh, B. J., & Rhodes, C. J. (2007). The balance between proinsulin biosynthesis and insulin secretion: where can imbalance lead?. Diabetes, obesity & metabolism , 9 Suppl 2, 56–66. https://doi.org/10.1111/j.1463-1326.2007.00774.x

15- Reily, C., Stewart, T. J., Renfrow, M. B., & Novak, J. (2019). Glycosylation in health and disease. Nature reviews. Nephrology, 15(6), 346–366. https://doi.org/10.1038/s41581-019-0129-4

16- Omar-Hmeadi, M., & Idevall-Hagren, O. (2021). Insulin granule biogenesis and exocytosis . Cellular and molecular life sciences : CMLS, 78(5), 1957–1970. https://doi.org/10.1007/s00018-020-03688-4

17- Thota, S., & Akbar, A. (2023). Insulin. In StatPearls. StatPearls Publishing.

18- Weiss, M., Steiner, D. F., & Philipson, L. H. (2014) . Insulin Biosynthesis, Secretion , Structure , and Structure-Activity Relationships. In K. R. Feingold (Eds.) et . al., Endotext. MDText.com, Inc.

19- Ramazi, S., & Zahiri, J. (2021). Posttranslational modifications in proteins: resources, tools and prediction methods. Database : the journal of biological databases and curation, 2021, baab012. https://doi.org/10.1093/database/baab012

20- Tokarz, V. L., MacDonald, P. E., & Klip, A. (2018). The cell biology of systemic insulin function . The Journal of cell biology, 217(7),2273–2289. https://doi.org/10.1083/jcb.201802095

21- Cohen P. (2002). The origins of protein phosphorylation. Nature cell biology, 4(5), E127–E130. https://doi.org/10.1038/ncb0502-e127

22- Rassier, D. E., & Månsson, A. (2025). Mechanisms of myosin II force generation: insights from novel experimental techniques and approaches. Physiological Reviews, 105(1), 1-93. doi:10.1152/physrev.00014.2023

23- Vargas, E., Joy, N. V., & Carrillo Sepulveda, M. A. (2022). Biochemistry, Insulin Metabolic Effects. In StatPearls. StatPearls Publishing.

24- Fiveable. (2024, August 1). Simple Proteins – Principles of Food Science. https://fiveable.me/key-terms/principles-food-science/simple-proteins.

25- Calvez, J., Azzout-Marniche, D., & Tomé, D. (2024). Protein quality, nutrition and health . Frontiers in nutrition , 11, 1406618. https://doi.org/10.3389/fnut.2024.1406618

26- Zhong, Q., Xiao, X., Qiu, Y., Xu, Z., Chen, C., Chong, B., Zhao, X., Hai, S., Li, S., An, Z., & Dai, L. (2023). Protein posttranslational modifications in health and diseases: Functions, regulatory mechanisms, and therapeutic implications. MedComm, 4(3), e261. https://doi.org/10.1002/mco2.261

27- Patricia, J.J. and Dhamoon, A.S. (2021) Physiology, Digestion. StatPearls Publishing, Treasure Island.

28- Cooper GM. The Cell: A Molecular Approach. 2nd edition. Sunderland (MA): Sinauer Associates; 2000. Translation of mRNA. Available from: https://www.ncbi.nlm.nih.gov/books/NBK9849/

29- Sun, Q., & Zhao, Z. (2017). Peptide Hormones as Tumor Markers in Clinical Practice . The Enzymes, 42, 65–79. https://doi.org/10.1016/bs.enz.2017.09.001

30- Rix, I., Nexøe-Larsen, C., Bergmann, N. C., Lund, A., & Knop, F. K. (2000). Glucagon physiology. In Endotext . MDText.com, Inc.

31- Wu G. (2009). Amino acids: metabolism, functions, and nutrition. Amino acids, 37(1), 1–17. https://doi.org/10.1007/s00726-009-0269-0

32- Chandel N. S. (2021). Amino Acid Metabolism. Cold Spring Harbor perspectives in biology, 13(4), a040584. https://doi.org/10.1101/cshperspect.a040584

33- Hoerter, J. E., & Ellis, S. R. (2023). Biochemistry, Protein Synthesis. In StatPearls. StatPearls Publishing.

34- Germanos, M., Gao, A., Taper, M., Yau, B., & Kebede, M. A. (2021). Inside the Insulin Secretory Granule. Metabolites, 11(8), 515. https://doi.org/10.3390/metabo11080515

35- Lönnerdal, B., Erdmann, P., Thakkar, S. K., Sauser, J., & Destaillats, F. (2017). Longitudinal evolution of true protein, amino acids and bioactive proteins in breast milk: a developmental perspective. The Journal of nutritional biochemistry, 41, 1–11. https://doi.org/10.1016/j.jnutbio.2016.06.001

36- Gremese, E., Bruno, D., Varriano, V., Perniola, S., Petricca, L., & Ferraccioli, G. (2023). Serum Albumin Levels: A Biomarker to Be Repurposed in Different Disease Settings in Clinical Practice. Journal of Clinical Medicine , 12(18), 6017. https://doi.org/10.3390/jcm12186017

37- Taverna, M., Marie, A. L., Mira, J. P., & Guidet, B. (2013). Specific antioxidant properties of human serum albumin . Annals of intensive care , 3(1), 4. https://doi.org/10.1186/2110-5820-3-4

38- Moman, R. N., Gupta, N., & Varacallo, M. A. (2022). Physiology, Albumin. In StatPearls. StatPearls Publishing.

39- Morris, C. H., & Baker, J. (2025). Glucagon. In StatPearls. StatPearls Publishing.

40- Li, S., Cao, Y., & Geng, F. (2017). Genome-Wide Identification and Comparative Analysis of Albumin Family in Vertebrates . Evolutionary bioinformatics online, 13, 1176934317716089 . https://doi.org/10.1177/1176934317716089

41- Anguizola, J., Matsuda, R., Barnaby, O. S., Hoy, K. S., Wa, C., DeBolt, E., Koke, M., & Hage, D. S. (2013). Review: Glycation of human serum albumin. Clinica chimica acta; international journal of clinical chemistry, 425, 64–76. https://doi.org/10.1016/j.cca.2013.07.013

42- Joshi, J. G., & Clauberg, M. (1988). Ferritin: an iron storage protein with diverse functions. BioFactors (Oxford, England), 1(3), 207–212.

43- Pollard, T. D., & Weihing, R. R. (1974). Actin and myosin and cell movement. CRC critical reviews in biochemistry, 2(1), 1–65. https://doi.org/10.3109/10409237409105443

44- Khalil, B., Marwaha, K., & Bollu, P. C. (2025). Physiology, Neuromuscular Junction . In StatPearls . StatPearls Publishing.

45- Clarke, M., & Spudich, J. A. (1977). Nonmuscle contractile proteins: the role of actin and myosin in cell motility and shape determination. Annual review of biochemistry, 46, 797–822 . https://doi.org/10.1146/annurev.bi.46.070177.004053

46- Lowey, S., & Steiner, L. A. (1972). An immunochemical approach to the structure of myosin and the thick filament . Journal of molecular biology, 65(1), 111–126 . https://doi.org/10.1016/0022-2836(72)90495-0

47- Korn E. D. (1978). Biochemistry of actomyosin-dependent cell motility (a review) . Proceedings of the National Academy of Sciences of the United States of America, 75(2), 588–599. https://doi.org/10.1073/pnas.75.2.588

48- Wang, K., McClure, J., & Tu, A. (1979). Titin: major myofibrillar components of striated muscle. Proceedings of the National Academy of Sciences of the United States of America, 76(8), 3698–3702 . https://doi.org/10.1073/pnas.76.8.3698

49- LeWinter, M. M., Wu, Y., Labeit, S., & Granzier, H. (2007). Cardiac titin: structure, functions and role in disease. Clinica chimica acta; international journal of clinical chemistry, 375(1-2), 1–9 . https://doi.org/10.1016/j.cca.2006.06.035

50- Lagúnez-Otero, J., Díaz-Villaseñor, A., & Renugopalakrishnan, V. (2002). Specialized biology from tandem beta-turns . Archives of medical research, 33(3), 245–249 . https://doi.org/10.1016/s0188-4409(02)00355-7

51- Alix A. J. (2001). Un tournant essentiel dans la connaissance des relations structure--fonction--activité de l'élastine [A turning point in the knowledge of the structure-function-activity relations of elastin]. Journal de la Societe de biologie, 195(2), 181–193 .

52- Granzier, H., Helmes, M., Cazorla, O., McNabb, M., Labeit, D., Wu, Y., Yamasaki, R., Redkar, A., Kellermayer, M., Labeit, S., & Trombitás, K. (2000). Mechanical properties of titin isoforms. Advances in experimental medicine and biology, 481, 283–304 . https://doi.org/10.1007/978-1-4615-4267-4_17

53- Noto, R. E., Leavitt, L., & Edens, M. A. (2023). Physiology, Muscle. In StatPearls. StatPearls Publishing.

54- Mishra, V., & Heath, R. J. (2021). Structural and Biochemical Features of Human Serum Albumin Essential for Eukaryotic Cell Culture.

International journal of molecular sciences, 22(16), 8411. https://doi.org/10.3390/ijms22168411

55- Krüger, M., & Linke, W. A. (2011). The giant protein titin: a regulatory node that integrates myocyte signaling pathways. The Journal of biological chemistry, 286(12), 9905–9912 . https://doi.org/10.1074/jbc.R110.173260

56- Moll, R., Divo, M., & Langbein, L. (2008). The human keratins: biology and pathology. Histochemistry and cell biology, 129(6), 705–733 . https://doi.org/10.1007/s00418-008-0435-6

57- Chamcheu, J. C., Siddiqui, I. A., Syed, D. N., Adhami, V. M., Liovic, M., & Mukhtar, H. (2011). Keratin gene mutations in disorders of human skin and its appendages . Archives of biochemistry and biophysics, 508(2), 123–137 . https://doi.org/10.1016/j.abb.2010.12.019

58- Fu, Z., Gilbert, E. R., & Liu, D. (2013). Regulation of insulin synthesis and secretion and pancreatic Beta-cell dysfunction in diabetes. Current diabetes reviews, 9(1), 25–53.

59- Bellingham, C. M., Lillie, M. A., Gosline, J. M., Wright, G. M., Starcher, B. C., Bailey, A. J., Woodhouse, K. A., & Keeley, F. W. (2003). Recombinant human elastin polypeptides self-assemble into biomaterials with elastin-like properties. Biopolymers, 70(4), 445–455. https://doi.org/10.1002/bip.10512

60- Dominguez, R., & Holmes, K. C. (2011). Actin structure and function . Annual review of biophysics, 40, 169–186 . https://doi.org/10.1146/annurev-biophys-042910-155359

61- van Beek, E. M., Cochrane, F., Barclay, A. N., & van den Berg, T. K. (2005). Signal regulatory proteins in the immune system. Journal of immunology (Baltimore, Md. : 1950), 175(12), 7781–7787. https://doi.org/10.4049/jimmunol.175.12.7781

62- Suresh, R., & Diaz, R. J. (2021). The remodelling of actin composition as a hallmark of cancer . Translational oncology, 14(6), 101051 . https://doi.org/10.1016/j.tranon.2021.101051

63- Wilcox G. (2005). Insulin and insulin resistance. The Clinical biochemist. Reviews, 26(2), 19–39.

64- Pollard T. D. (2016). Actin and Actin-Binding Proteins. Cold Spring Harbor perspectives in biology, 8(8), a018226 . https://doi.org/10.1101/cshperspect.a018226

65- Wu, M., Cronin, K., & Crane, J. S. (2023). Biochemistry, Collagen Synthesis. In StatPearls. StatPearls Publishing.

66- Alberts, B., Johnson, A., Lewis, J., Raff, M., Roberts, K., & Walter, P. (2002). The Shape and Structure of Proteins. https://www.ncbi.nlm.nih.gov/books/NBK26830/

67- Hayward A. R. (1983). The human fetus and newborn: development of the immune response. Birth defects original article series, 19(3), 289–294.

68- Neurath, H. (1963). The proteins : composition, structure, and function. Volume I (2nd ed.). Academic Press.

69-Cozzone, A. J. (2010). Proteins: Fundamental chemical properties. eLS.

70- Whitford, D., & VLeBooks. (2005). Proteins : structure and function / David Whitford. J. Wiley & Sons.

https://www.vlebooks.com/vleweb/product/openreader?id=StAndrews&isbn=9780470012413

71- Kessel, A., & Ben-Tal, N. (2011). Introduction to proteins : structure, function, and motion. CRC Press.

72- Van Troys, M., Vandekerckhove, J., & Ampe, C. (1999). Structural modules in actin-binding proteins: towards a new classification. Biochimica et biophysica acta, 1448(3), 323–348. https://doi.org/10.1016/s0167-4889(98)00152-9

73- Gutzeit, C., Chen, K., & Cerutti, A. (2018). The enigmatic function of IgD: some answers at last. European journal of immunology, 48(7), 1101–1113. https://doi.org/10.1002/eji.201646547

74- Greenfield, E. A. (2014). Antibodies : a laboratory manual (Second edition.). Cold Spring Harbor Laboratory Press.

75- Forthal D. N. (2014). Functions of Antibodies. Microbiology spectrum, 2(4), . https://doi.org/10.1128/microbiolspec.AID-0019-2014.

76- Baumann, L., Bernstein, E. F., Weiss, A. S., Bates, D., Humphrey, S., Silberberg, M., & Daniels, R. (2021). Clinical Relevance of Elastin in the Structure and Function of Skin. Aesthetic surgery journal. Open forum, 3(3), ojab019. https://doi.org/10.1093/asjof/ojab019

77- Moerman, D. G., & Fire, A. (1997). Muscle: Structure, Function, and Development. In D. L. Riddle (Eds.) et. al., C. elegans II. (2nd ed.). Cold Spring Harbor Laboratory Press.

78- Chen, I., & Lui, F. (2022). Physiology, Active Transport. In StatPearls. StatPearls Publishing.

79- Litvinov, R. I., & Weisel, J. W. (2016). What Is the Biological and Clinical Relevance of Fibrin?. Seminars in thrombosis and hemostasis, 42(4), 333–343. https://doi.org/10.1055/s-0036-1571342

80- Owen, J. A., Punt, J., Stranford, S. A., Jones, P. P., & Kuby, J. (2013). Kuby immunology (Seventh edition.). W.H. Freeman.

81- Lydyard, P.M. (2003). BIOS Instant Notes in Immunology (2nd ed.). Taylor & Francis. https://doi.org/10.4324/9780203488287

82- Bailey, A., & Mohiuddin, S. S. (2022). Biochemistry, High Density Lipoprotein. In StatPearls. StatPearls Publishing.

83- Patel, A., & Jialal, I. (2023). Biochemistry, Immunoglobulin A. In StatPearls. StatPearls Publishing.

84- Sutton, B. J., Davies, A. M., Bax, H. J., & Karagiannis, S. N. (2019). IgE Antibodies: From Structure to Function and Clinical Translation. Antibodies (Basel, Switzerland), 8(1), 19 . https://doi.org/10.3390/antib8010019

85- Balla, J., Rathore, A. P. S., & St John, A. L. (2025). Maternal IgE Influence on Fetal and Infant Health . Immunological reviews, 331(1), e70029. https://doi.org/10.1111/imr.70029

86- Vidarsson, G., Dekkers, G., & Rispens, T. (2014) . IgG subclasses and allotypes: from structure to effector functions. Frontiers in immunology, 5, 520 . https://doi.org/10.3389/fimmu.2014.00520

87- Patel, P., Jamal, Z., & Ramphul, K. (2023). Immunoglobulin. In StatPearls. StatPearls Publishing.

88- Aziz, M., Iheanacho, F., & Hashmi, M. F. (2025). Physiology, antibody . In StatPearls. StatPearls Publishing.

89- Zhang, J. M., & An, J. (2007). Cytokines, inflammation, and pain. International anesthesiology clinics, 45(2), 27–37 . https://doi.org/10.1097/AIA.0b013e318034194e

90- Lewis, T., & Stone, W. L. (2023). Biochemistry, Proteins Enzymes. In StatPearls. StatPearls Publishing.

91- Kessler, A. T., & Raja, A. (2023). Biochemistry, Histidine. In StatPearls. StatPearls Publishing.

92- Lopez, M.J. and Mohiuddin, S.S. (2021) Biochemistry, Essential Amino Acids. StatPearls Publishing, Treasure Island.

93- Cameron, M. J., & Kelvin, D. J. (2003). Cytokines and chemokines--their receptors and their genes: an overview. Advances in experimental medicine and biology, 520, 8–32. https://doi.org/10.1007/978-1-4615-0171-8_2.

94- Vanek, T., & Kohli, A. (2023). Biochemistry, Myoglobin. In StatPearls. StatPearls Publishing.

95- Ogun, A. S., & Adeyinka, A. (2025). Biochemistry, transferrin. In StatPearls. StatPearls Publishing.

96- Ahmed, M. H., Ghatge, M. S., & Safo, M. K. (2020). Hemoglobin: Structure, Function and Allostery. Sub-cellular biochemistry, 94, 345–382. https://doi.org/10.1007/978-3-030-41769-7_14

97- Billett, H. H. (1990). Hemoglobin and Hematocrit. In H. K. Walker (Eds.) et. al., Clinical Methods: The History, Physical, and Laboratory Examinations. (3rd ed.). Butterworths.

98- Zhang, Y., & Zheng, J. (2020). Bioinformatics of Metalloproteins and Metalloproteomes. Molecules (Basel, Switzerland), 25(15), 3366. https://doi.org/10.3390/molecules25153366

99- Lu, Y., Yeung, N., Sieracki, N., & Marshall, N. M. (2009). Design of functional metalloproteins. Nature, 460(7257), 855–862. https://doi.org/10.1038/nature08304

100- Venerando, A., Cesaro, L., & Pinna, L. A. (2017). From phosphoproteins to phosphoproteomes: a historical account. The FEBS journal, 284(13), 1936–1951. https://doi.org/10.1111/febs.14014

101- Watford, M., & Wu, G. (2018). Protein. Advances in nutrition (Bethesda, Md.), 9(5), 651–653. https://doi.org/10.1093/advances/nmy027

102- Jiang, B., Wang, L., Wang, X., Wu, S., Li, D., Liu, C., & Feng, Z. (2019). Ultrasonic Thermal-Assisted Extraction of Phosvitin from Egg Yolk and Evaluation of Its Properties. Polymers, 11(8), 1353. https://doi.org/10.3390/polym11081353

103- Ferris, S. P., Kodali, V. K., & Kaufman, R. J. (2014). Glycoprotein folding and quality-control mechanisms in protein-folding diseases. Disease models & mechanisms, 7(3), 331–341. https://doi.org/10.1242/dmm.014589

104- Vladutiu A. O. (2000). Immunoglobulin D: properties, measurement, and clinical relevance. Clinical and diagnostic laboratory immunology, 7(2), 131–140. https://doi.org/10.1128/CDLI.7.2.131-140.2000

105- Franczyk, B., Gluba-Brzózka, A., Ciałkowska-Rysz, A., Ławiński, J., & Rysz, J. (2023). The Impact of Aerobic Exercise on HDL Quantity and Quality: A Narrative Review. International journal of molecular sciences, 24(5), 4653. https://doi.org/10.3390/ijms24054653

106- Huang, S., Li, J., Shearer, G. C., Lichtenstein, A. H., Zheng, X., Wu, Y., Jin, C., Wu, S., & Gao, X. (2017). Longitudinal study of alcohol consumption and HDL concentrations: a community-based study. The American journal of clinical nutrition, 105(4), 905–912. https://doi.org/10.3945/ajcn.116.144832

107- Narang, D., & Al-Horani, R. A. (2025). Biochemistry, intermediate density lipoprotein. In StatPearls. StatPearls Publishing.

108- Pirahanchi, Y., Sinawe, H., & Dimri, M. (2023). Biochemistry, LDL Cholesterol. In StatPearls. StatPearls Publishing.

109- Rahmany, S., & Jialal, I. (2023). Biochemistry, Chylomicron. In StatPearls. StatPearls Publishing.

110- Venugopal, S. K., Anoruo, M., & Jialal, I. (2023). Biochemistry, Low Density Lipoprotein. In StatPearls. StatPearls Publishing.

111- Feingold, K. R. (2024). Introduction to Lipids and Lipoproteins. In K. R. Feingold (Eds.) et. al., Endotext. MDText.com, Inc.

112- Farzam, K., Zubair, M., & Senthilkumaran, S. (2024). Lipoprotein A. In StatPearls. StatPearls Publishing.

113- Lee, H. C., Akhmedov, A., & Chen, C. H. (2022). Spotlight on very-low-density lipoprotein as a driver of cardiometabolic disorders: Implications for disease progression and mechanistic insights. Frontiers in cardiovascular medicine, 9, 993633. https://doi.org/10.3389/fcvm.2022.993633

114- Juarez Casso, F. M., & Farzam, K. (2022). Biochemistry, Very Low Density Lipoprotein. In StatPearls. StatPearls Publishing.

115- Zeng, B. J., Mortimer, B. C., Martins, I. J., Seydel, U., & Redgrave, T. G. (1998). Chylomicron remnant uptake is regulated by the expression and function of heparan sulfate proteoglycan in hepatocytes. Journal of lipid research, 39(4), 845–860.

116- de Vries, J. J., Snoek, C. J. M., Rijken, D. C., & de Maat, M. P. M. (2020). Effects of Post-Translational Modifications of Fibrinogen on Clot Formation, Clot Structure, and Fibrinolysis: A Systematic Review. Arteriosclerosis, thrombosis, and vascular biology, 40(3), 554–569. https://doi.org/10.1161/ATVBAHA.119.313626

118- Kang, T. S., & Kini, R. M. (2009). Structural determinants of protein folding. Cellular and molecular life sciences : CMLS, 66(14), 2341–2361. https://doi.org/10.1007/s00018-009-0023-5

119- Hartman, M. A., & Spudich, J. A. (2012). The myosin superfamily at a glance. Journal of cell science, 125(Pt 7), 1627–1632. https://doi.org/10.1242/jcs.094300

120- Mosesson M. W. (2005). Fibrinogen and fibrin structure and functions. Journal of thrombosis and haemostasis : JTH, 3(8), 1894–1904. https://doi.org/10.1111/j.1538-7836.2005.01365.x

121- Cooper, G., & Adams, K. (2023, August 31). Protein Synthesis, Processing, and Regulation. Science Trove. Retrieved 29 Nov. 2025, from https://www.oxfordsciencetrove.com/view/10.1093/hesc/9780197583746.001.0001/isbn-9780197583746-book-part-10.

FSC
www.fsc.org
MIX
Papier | Fördert gute Waldnutzung
FSC® C083411

Kolibri 360 GmbH
Ferdinand-Jühlke-Straße 7
99095 Erfurt, Deutschland
produktsicherheit@kolibri360.de

Druck:
CPI Druckdienstleistungen GmbH
im Auftrag der
Kolibri 360 GmbH
Ferdinand-Jühlke-Str. 7
99095 Erfurt